Also by Paul Davies

The Runaway Universe
Other Worlds
The Edge of Infinity
God and the New Physics
Superforce

Student Texts

Space and time in the modern universe
The forces of nature
The search for gravity waves
The accidental universe
Quantum mechanics

Technical

The physics of time asymmetry
Quantum fields in curved space (with N. D. Birrell)

The

C·O·S·M·I·C

Blueprint

New Discoveries
in Nature's Creative Ability
to Order the Universe

Paul Davies

A TOUCHSTONE BOOK
Published by Simon & Schuster Inc.
NEW YORK • LONDON • TORONTO • SYDNEY • TOKYO

Touchstone
Simon & Schuster Building
Rockefeller Center
1230 Avenue of the Americas
New York, New York 10020

First Touchstone Edition, 1989

TOUCHSTONE and colophon are registered trademarks
of Simon & Schuster Inc.

Manufactured in the United States of America

1   3   5   7   9   10   8   6   4   2
1   3   5   7   9   10   8   6   4   2 Pbk.

Library of Congress Cataloging in Publication data

Davies, P. C. W.
The cosmic blueprint.

Includes index.
1. Force and energy. 2. Matter.
3. Self-organizing systems. 4. Cosmology.
I. Title.
QC73.D383   1988   530   87-26442

ISBN 0-671-60233-0
0-671-67561-3 Pbk.

Whether or not it is clear to you, no doubt the universe is unfolding as it should.

Max Ehrmann

Contents

Contents

The Cosmic Blueprint

Preface

The creation of the universe is usually envisaged as an abrupt event that took place in the remote past. It is a picture reinforced both by religion and by scientific evidence for a 'big bang'. What this simple idea conceals, however, is that the universe has never ceased to be creative.

Cosmologists now believe that immediately following the big bang the universe was in an essentially featureless state, and that all the structure and complexity of the physical world we see today somehow emerged afterwards. Evidently physical processes exist that can turn a void – or something close to it – into stars, planets, crystals, clouds and people.

What is the source of this astonishing creative power? Can known physical processes explain the continuing creativity of nature, or are there additional organizing principles at work, shaping matter and energy and directing them towards ever higher states of order and complexity?

Only very recently have scientists begun to understand how complexity and organization can emerge from featurelessness and chaos. Research in areas as diverse as fluid turbulence, crystal growth and neural networks is revealing the extraordinary propensity for physical systems to generate new states of order spontaneously. It is clear that there exist *self-organizing* processes in every branch of science.

A fundamental question then presents itself. Are the seemingly endless varieties of natural forms and structures, which appear as the universe unfolds, simply the accidental products of random forces? Or are they somehow the inevitable outcome of the creative activity of nature? The origin of life, for example, is regarded by some scientists as an extremely rare chance event, but by others as the natural end state of cycles of self-organizing chemical reactions. If the richness of nature is built into its laws, does this imply that the present state of the universe is in some sense predestined? Is there, to use a metaphor, a 'cosmic blueprint'?

These deep questions of existence are not, of course, new. They have been asked by philosophers and theologians for millenia. What makes them especially germane today is that important new discoveries are dramatically altering the *scientists'* perspective of the nature of the universe. For three centuries science has been dominated by the Newtonian and thermodynamic paradigms, which present the universe either as a sterile machine, or in a state of degeneration and decay. Now there is the new paradigm of the creative universe, which recognizes the progressive, innovative character of physical processes. The new paradigm emphasizes the collective, cooperative and organizational aspects of nature; its perspective is synthetic and holistic rather than analytic and reductionist.

This book is an attempt to bring these significant developments to the attention of the general reader. It covers new research in many disciplines, from astronomy to biology, from physics to neurology – wherever complexity and self-organization appear. I have tried to keep the presentation as non-technical as possible, but inevitably there are some key sections that require a more careful treatment. This is especially true of Chapter 4, which contains a number of technical diagrams. The reader is urged to persevere, however, for the essence of the new paradigm cannot be properly captured without some mathematical ideas.

In compiling the material I have been greatly assisted by my colleagues at the University of Newcastle upon Tyne, who do not, of course, necessarily share my conclusions. Particular thanks are due to Professor Kenneth Burton, Dr Ian Moss, Dr Richard Rohwer and Dr David Tritton. I should like to thank Dr John Barrow, Professor Roger Penrose and Professor Frank Tipler for helpful discussions.

1988   P.D.

Since this book went to press there have been a number of interesting publications relevant to the topic of complexity. Especially pertinent are the ideas of physicist Anthony J. Leggett, who conjectures (as do I) that new physical effects might emerge at some particular scale of complexity. His book, *The Problems of Physics* (Oxford University Press, 1988), is highly recommended. Very exciting work on the topic of algorithmic complexity theory as applied to physical systems is being carried out by Gregory J. Chaitin of IBM. Some of this is published in his book, *Algorithmic Information Theory* (Cambridge University Press, 1987). I am indebted to Dr. Chaitin for some very stimulating correspondence.

1989   P.D.

I

Blueprint for a Universe

God is no more an archivist unfolding an infinite sequence he had designed once and forever. He continues the labour of creation throughout time.

Ilya Prigogine[1]

The origin of things

Something buried deep in the human psyche compels us to contemplate creation. It is obvious even at a casual glance that the universe is remarkably ordered on all scales. Matter and energy are distributed neither uniformly nor haphazardly, but are organized into coherent identifiable structures, occasionally of great complexity. From whence came the myriads of galaxies, stars and planets, the crystals and clouds, the living organisms? How have they been arranged in such harmonious and ingenious interdependence? The cosmos, its awesome immensity, its rich diversity of forms, and above all its coherent unity, cannot be accepted simply as a brute fact.

The existence of *complex* things is even more remarkable given the generally delicate and specific nature of their organization, for they are continually assailed by all manner of disruptive influences from their environment that care nothing for their survival. Yet in the face of an apparently callous Mother Nature the orderly arrangement of the universe not only manages to survive, but to prosper.

There have always been those who choose to interpret the harmony and order of the cosmos as evidence for a metaphysical designer. For them, the existence of complex forms is explained as a manifestation of the

3

designer's creative power. The rise of modern science, however, transformed the rational approach to the problem of the origin of things. It was discovered that the universe has not always been as it is. The evidence of geology, palaeontology and astronomy suggested that the vast array of forms and structures that populate our world have not always existed, but have *emerged* over aeons of time.

Scientists have recently come to realize that *none* of the objects and systems that make up the physical world we now perceive existed in the beginning. Somehow, all the variety and complexity of the universe has arisen since its origin in an abrupt outburst called the big bang. The modern picture of Genesis is of a cosmos starting out in an utterly featureless state, and then progressing step by step – one may say *unfolding* – to the present kaleidoscope of organized activity.

Creation from nothing

The philosopher Parmenides, who lived 1500 years before Christ, taught that 'nothing can come out of nothing'. It is a dictum that has been echoed many times since, and it forms the basis of the approach to creation in many of the world's religions, such as Judaism and Christianity. Parmenides' followers went much farther, to conclude that there can be no real change in the physical world. All apparent change, they asserted, is an illusion. Theirs is a dismally sterile universe, incapable of bringing forth anything fundamentally new.

Believers in Parmenides' dictum cannot accept that the universe came into existence spontaneously; it must either always have existed or else have been created by a supernatural power. The Bible states explicitly that God created the world, and Christian theologians advance the idea of creation *ex nihilo* – out of literally nothing. Only God, it is said, possesses the power to accomplish this.

The problem of the ultimate origin of the physical universe lies on the boundary of science. Indeed, many scientists would say it lies beyond the scope of science altogether. Nevertheless, there have recently been serious attempts to understand how the universe could have appeared from nothing without violating any physical laws. But how can something come into existence uncaused?

The key to achieving this seeming miracle is *quantum physics*. Quantum processes are *inherently* unpredictable and indeterministic; it is generally impossible to predict from one moment to the next how a quantum system

4

will behave. The law of cause and effect, so solidly rooted in the ground of daily experience, fails here. In the world of the quantum, spontaneous change is not only permitted, it is unavoidable.

Although quantum effects are normally restricted to the microworld of atoms and their constituents, in principle quantum physics should apply to everything. It has become fashionable to investigate the quantum physics of the entire universe, a subject known as quantum cosmology. These investigations are tentative and extremely speculative, but they lead to a provocative possibility. It is no longer entirely absurd to imagine that the universe came into existence spontaneously from nothing as a result of a quantum process.

The fact that the nascent cosmos was apparently devoid of form and content greatly eases the problem of its ultimate origin. It is much easier to believe that a state of featureless simplicity appeared spontaneously out of nothing than to believe that the present highly complex state of the universe just popped into existence ready-made.

The amelioration of one problem, however, leads immediately to another. Science is now faced with the task of explaining by what physical processes the organized systems and elaborate activity that surround us today emerged from the primeval blandness of the big bang. Having found a way of permitting the universe to be self-creating we need to attribute to it the capability of being *self-organizing*.

An increasing number of scientists and writers have come to realize that the ability of the physical world to organize itself constitutes a fundamental, and deeply mysterious, property of the universe. The fact that nature has *creative power*, and is able to produce a progressively richer variety of complex forms and structures, challenges the very foundation of contemporary science. 'The greatest riddle of cosmology,' writes Karl Popper, the well-known philosopher, 'may well be ... that the universe is, in a sense, creative.'[2]

The Belgian Nobel prize-winner Ilya Prigogine, writing with Isabelle Stengers in their book *Order Out of Chaos*, reaches similar conclusions:[3] 'Our universe has a pluralistic, complex character. Structures may disappear, but also they may appear.' Prigogine and Stengers dedicate their book to Erich Jantsch, whose earlier work *The Self-Organizing Universe* also expounds the view that nature has a sort of 'free will' and is thereby capable of generating novelty:[4] 'We may one day perhaps understand the self-organizing processes of a universe which is not determined by the blind selection of initial conditions, but has the potential of partial self-determination.'

These sweeping new ideas have not escaped the attention of the science writers. Louise Young, for example, in lyrical style, refers to the universe as 'unfinished', and elaborates Popper's theme:[5] 'I postulate that we are witnessing – and indeed participating in – a creative act that is taking place throughout time. As in all such endeavours, the finished product could not have been clearly foreseen in the beginning.' She compares the unfolding organization of the cosmos with the creative act of an artist: '. . . involving change and growth, it proceeds by trial and error, rejecting and reformulating the materials at hand as new potentialities emerge'.

In recent years much attention has been given to the problem of the so-called 'origin of the universe', and popular science books on 'the creation' abound. The impression is gained that the universe was created all at once in the big bang. It is becoming increasingly clear, however, that creation is really a continuing process. The existence of the universe is not explained by the big bang: the primeval explosion merely started things off.

Now we must ask: How can the universe, having come into being, subsequently bring into existence totally new things by following the laws of nature? Put another way: What is the source of the universe's creative potency? It will be the central question of this book.

The whole and its parts

To most people it is obvious that the universe forms a coherent whole. We recognize that there are a great many components that go together to make up the totality of existence, but they seem to hang together, if not in cooperation, then at least in peaceful coexistence. In short, we find order, unity and harmony in nature where there might have been discord and chaos.

The Greek philosopher Aristotle constructed a picture of the universe closely in accord with this intuitive feeling of holistic harmony. Central to Aristotle's philosophy was the concept of *teleology* or, roughly speaking, final causation. He supposed that individual objects and systems subordinate their behaviour to an overall plan or destiny. This is especially apparent, he claimed, in living systems, where the component parts function in a cooperative way to achieve a final purpose or end product. Aristotle believed that living organisms behave as a coherent whole because there exists a full and perfect 'idea' of the entire organism, even before it develops. The development and behaviour of living things is thus guided and controlled by the global plan in order that it should successfully approach its designated end.

Aristotle extended this animistic philosophy to the cosmos as a whole. There exists, he maintained, what we might today term a *cosmic blueprint*. The universe was regarded as a sort of gigantic organism, unfurling in a systematic and supervised way towards its prescribed destiny. Aristotelian finalism and teleology later found its way into Christian theology, and even today forms the basis of Western religious thought. According to Christian dogma, there is indeed a cosmic blueprint, representing God's design for a universe.

In direct opposition to Aristotle were the Greek atomists, such as Democritus, who taught that the world is nothing but atoms moving in a void. All structures and forms were regarded as merely different arrangements of atoms, and all change and process were thought of as due to the rearrangement of atoms alone. To the atomist, the universe is a machine in which each component atom moves entirely under the action of the blind forces produced by its neighbours. According to this scheme there are no final causes, no overall plan or end-state towards which things evolve. Teleology is dismissed as mystical. The only causes that bring about change are those produced by the shape and movement of other atoms.

Atomism is not suited to describe, let alone explain, the order and harmony of the world. Consider a living organism. It is hard to resist the impression that the atoms of the organism *cooperate* so that their collective behaviour constitutes a coherent unity. The organized functioning of biological systems fails to be captured by a description in which each atom is simply pushed or pulled along blindly by its neighbours, without reference to the global pattern. There was thus already present in ancient Greece the deep conflict between holism and reductionism which persists to this day. On the one hand stood Aristotle's synthetic, purposeful universe, and on the other a strictly materialistic world which could ultimately be analysed as, or reduced to, the simple mechanical activity of elementary particles.

In the centuries that followed, Democritus' atomism came to represent what we would now call the scientific approach to the world. Aristotelian ideas were banished from the physical sciences during the Renaissance. They survived somewhat longer in the biological sciences, if only because living organisms so distinctly display teleological behaviour. However, Darwin's theory of evolution and the rise of modern molecular biology led

7

to the emphatic rejection of all forms of animism or finalism, and most modern biologists are strongly mechanistic and reductionist in their approach. Living organisms are today generally regarded as purely complex machines, programmed at the molecular level.

The scientific paradigm in which all physical phenomena are reduced to the mechanical behaviour of their elementary constituents has proved extremely successful, and has led to many new and important discoveries. Yet there is a growing dissatisfaction with sweeping reductionism, a feeling that the whole really is greater than the sum of its parts. Analysis and reduction will always have a central role to play in science, but many people cannot accept that it is an exclusive role. Especially in physics, the synthetic or holistic approach is becoming increasingly fashionable in tackling certain types of problem.

However, even if one accepts the need to complement reductionism with a holistic account of nature, many scientists would still reject the idea of a cosmic blueprint as too mystical, for it implies that the universe has a purpose and is the product of a metaphysical designer. Such beliefs have been taboo for a long time among scientists. Perhaps the apparent unity of the universe is merely an anthropocentric projection. Or maybe the universe behaves *as if* it is implementing the design of a blueprint, but nevertheless is still evolving in blind conformity with purposeless laws?

These deep issues of existence have accompanied the advance of knowledge since the dawn of the scientific era. What makes them so pertinent today is the sweeping nature of recent discoveries in cosmology, fundamental physics and biology. In the coming chapters we shall see how scientists, in building up a picture of how organization and complexity arise in nature, are beginning to understand the origin of the universe's creative power.

2

The Missing Arrow

A clockwork universe

Every marksman knows that if a bullet misses its target, the gun was not aimed correctly. The statement seems trite, yet it conceals a deep truth. The fact that a bullet will follow a definite path in space from gun to target, and that this path is completely determined by the magnitude and direction of the muzzle velocity, is a clear example of what we might call the dependability of nature. The marksman, confident in the unfailing relationship between cause and effect, can estimate in advance the trajectory of the bullet. He will know that if the gun is accurately aligned the bullet will hit the target.

The marksman's confidence rests on that huge body of knowledge known as classical mechanics. Its origins stretch back into antiquity; every primitive hunter must have recognized that the flight of a stone from a sling or an arrow from a bow was not a haphazard affair, the main uncertainty being the act of projection itself. However, it was not until the seventeeth century, with the work of Galileo Galilei and Isaac Newton, that the laws of motion were properly formulated. In his monumental work *Principia*, published in 1687, Newton expounded his three famous laws that govern the motion of material bodies.

Cast in the form of mathematical equations, Newton's three laws imply that the motion of a body through space is determined entirely by the forces that act on the body, once its initial position and velocity are fixed. In the case of the bullet, the only significant force is the pull of gravity, which causes the path of the bullet to arch slightly into a parabolic curve.

Newton recognized that gravity also curves the paths of the planets around the Sun, in this case into ellipses. It was a great triumph that his laws of motion correctly described not only the shapes but also the periods of the planetary orbits. Thus was it demonstrated that even the heavenly bodies comply with universal laws of motion. Newton and his contemporaries were able to give an ever more accurate and detailed account of the workings of the solar system. The astronomer Halley, for example, computed the orbit of his famous comet, and was thereby able to give the date of its reappearance.

As the calculations became progressively more refined (and complicated) so the positions of planets, comets and asteroids could be predicted with growing precision. If a discrepancy appeared, then it could be traced to the effect of some contributing force that had been overlooked. The planets Uranus, Neptune and Pluto were discovered because their gravitational fields produced otherwise unaccountable perturbations in the orbits of the planets.

In spite of the fact that any given calculation could obviously be carried out to a finite accuracy only, there was a general assumption that the motion of every fragment of matter in the universe could in principle be computed to arbitrary precision if all the contributory forces were known. This assumption seemed to be spectacularly validated in astronomy, where gravity is the dominant force. It was much harder, however, to test in the case of smaller bodies subject to a wide range of poorly understood forces. Nevertheless Newton's laws were supposed to apply to *all* particles of matter, including individual atoms.

It came to be realized that a startling conclusion must follow. If every particle of matter is subject to Newton's laws, so that its motion is entirely determined by the initial conditions and the pattern of forces arising from all the other particles, then everything that happens in the universe, right down to the smallest movement of an atom, must be fixed in complete detail.

This arresting inference was made explicit in a famous statement by the French physicist Pierre Laplace:[1]

> Consider an intelligence which, at any instant, could have a knowledge of all forces controlling nature together with the momentary conditions of all the entities of which nature consists. If this intelligence were powerful enough to submit all this data to analysis it would be able to embrace in a single formula the movements of the largest bodies in the universe and those of the lightest atoms; for it nothing would be uncertain; the future and the past would be equally present to its eyes.

Laplace's claim implies that everything that has ever happened in the universe, everything that is happening now, and everything that ever will happen, has been unalterably determined from the first instant of time.

The future may be uncertain to our eyes, but it is *already fixed in every minute detail*. No human decisions or actions can change the fate of a single atom, for we too are part of the physical universe. However much we may feel free, everything that we do is, according to Laplace, completely determined. Indeed the entire cosmos is reduced to a gigantic clockwork mechanism, with each component slavishly and unfailingly executing its preprogrammed instructions to mathematical precision. Such is the sweeping implication of Newtonian mechanics.

Necessity

The determinism implicit in the Newtonian world view can be expressed by saying that every event happens *of necessity*. It *has* to happen; the universe has no choice. Let us take a closer look at how this necessity is formulated.

An essential feature of the Newtonian paradigm is that the world, or a part of it, can be ascribed a *state*. This state may be the position and velocity of a particle, the temperature and pressure of a gas or some more complicated set of quantities. When things happen in the world, the states of physical systems change. The Newtonian paradigm holds that these changes can be understood in terms of the forces that act on the system, in accordance with certain dynamical laws that are themselves independent of the states.

The success of the scientific method can be attributed in large measure to the ability of the scientist to discover universal laws which enable certain common features to be discerned in different physical systems. For example, bullets follow parabolic paths. If every system required its own individual description there would be no science as we know it. On the other hand the world would be dull indeed if the laws of motion alone were sufficient to fix everything that happens. In practice, the laws describe *classes* of behaviour. In any individual case they must be supplemented by specifying certain initial conditions. For example, the marksman needs to know the direction and velocity of the bullet at the muzzle before a unique parabolic trajectory is determined.

The interplay between states and dynamical laws is such that, given the laws, the state of a system at one moment *determines* its states at all subsequent moments. This element of determinism that Newton built into mechanics has grown to pervade all science. It forms the basis of scientific *testing*, by providing for the possibility of prediction.

The heart of the scientific method is the ability of the scientist to mirror or model events in the real world using mathematics. The theoretical physicist, for example, can set down the relevant dynamical laws in the form of equations, feed in the details about the initial state of the system he is modelling, and then solve the equations to find out how the system will evolve. The sequence of events that befalls the system in the real world is mirrored in the mathematics. In this way one may say that mathematics can mimic reality.

In choosing which equations to employ to describe the evolution of a physical system, note must be taken of certain requirements. One obvious property which the equations must possess is that, for all possible states of the system, a solution to the equations must exist. Furthermore that solution must be unique, otherwise the mathematics mimics more than one possible reality. The dual requirements of existence and uniqueness impose very strong restrictions on the form of the equations that can be used. In practice, the physicist usually uses second-order differential equations. The deterministic connection between sequences of physical states is paralleled in the mathematics by the logical dependence that various quantities in the equations have on one another. This is most obvious if a computer is solving the equations to simulate the evolution of some dynamical system. Each step of the computation is then logically determined by the previous step as the simulation proceeds.

In the three centuries that followed the publication of the *Principia*, physics underwent major convulsions, and the original Newtonian conception of the world has been enormously enlarged. Today, the truly fundamental material entities are no longer considered to be particles, but *fields*. Particles are regarded as disturbances in the fields, and so have been reduced to a derivative status. Nevertheless the fields are still treated according to the Newtonian paradigm, their activity determined by laws of motion plus initial conditions. Nor has the essence of the paradigm changed with the quantum and relativity revolutions that altered so profoundly our conception of space, time and matter. In all cases the system is still described in terms of states evolving deterministically according to fixed dynamical laws. Field or particle, everything that happens still happens 'of necessity'.

Reduction

The Newtonian paradigm fits in well with the philosophy of atomism discussed in the previous chapter. The behaviour of a macroscopic body can be reduced to the motion of its constituent atoms moving according to Newton's mechanistic laws. The procedure of breaking down physical systems into their elementary components and looking for an explanation of their behaviour at the lowest level is called *reductionism*, and it has exercised a powerful influence over scientific thinking.

So deeply has reductionism penetrated physics that the ultimate goal of the subject remains the identification of the fundamental fields (hence particles) and their dynamical behaviour in interaction. In recent years there has been spectacular progress towards this goal. Technically speaking, the aim of the theorist is to provide a mathematical expression known as a Lagrangian, after the French physicist Joseph Lagrange who provided an elegant formulation of Newton's laws. Given a Lagrangian for a system (whether consisting of fields, particles or both) there is a well-defined mathematical procedure for generating the dynamical equations from it.

A philosophy has grown up around this procedure that once a Lagrangian has been discovered that will accurately describe a system, then the behaviour of the system is considered to be 'explained'. In short, *a Lagrangian equals an explanation*. Thus, if a theorist could produce a Lagrangian that correctly accounts for all the observed fields and particles, nothing more is felt to be needed. If someone then asks for an explanation of the universe, in all its intricate complexity, the theoretical physicist would merely point to the Lagrangian and say: 'There! I've explained it all!'

This belief that all things ultimately flow from the fundamental Lagrangian goes almost unquestioned in the physics community. It has been succinctly expressed by Leon Lederman, director of the Fermi National Accelerator Laboratory near Chicago:[2] 'We hope to explain the entire universe in a single, simple formula [i.e. Lagrangian] that you can wear on your T-shirt.'

Not so long ago the Cambridge theorist Stephen Hawking took a similar line in his inaugural lecture as Lucasian Professor. As perhaps befits an incumbent of the chair once held by Newton, Hawking conjectured freely about the final triumph of the Newtonian paradigm. Excited by the rapid progress towards uncovering the fundamental Lagrangian of all known fields via an approach known as supergravity, Hawking entitled his lecture 'Is the end in sight for theoretical physics?' The implication, of course, was that given such a Lagrangian, theoretical physics would have reached its culmination, leaving only technical elaborations. The world would be 'explained'.

Whatever happened to time?

If the future is completely determined by the present, then in some sense the future is already contained in the present. The universe can be assigned a present state, which contains all the information needed to build the future – and by inversion of the argument, the past too. All of existence is thus somehow encapsulated, frozen, in a single instant. Time exists merely as a parameter for gauging the interval between events. Past and future have no real significance. Nothing actually *happens*.

Prigogine has called time 'the forgotten dimension' because of the impotence assigned to it by the Newtonian world view. In our ordinary experience time is not at all like that. Subjectively we feel that the world is changing, evolving. Past and future have distinct – and distinctive – meanings. The world appears to us as a movie. There is activity; things happen; time *flows*.

This subjective view of an active, evolving world is buttressed by observation. The changes that occur around us amount to more than Democritus' mere rearrangement of atoms in a void. True, atoms are rearranged, but in a systematic way that distinguishes past from future. It is only necessary to play a movie backwards to see the many everyday physical processes that are asymmetric in time. And not only in our own immediate experience. The universe as a whole is engaged in *unidirectional change*, an asymmetry often symbolized by an imaginary 'arrow of time', pointing from past to future.

How can these two divergent views of time be reconciled?

Newtonian time derives from a very basic property of the laws of motion: they are reversible. That is, the laws do not distinguish 'time forwards' from 'time backwards'; the arrow of time can point either way. From the standpoint of these laws, a movie played in reverse would be a perfectly acceptable sequence of real events. But from our point of view such a reversed sequence is impossible because most physical processes that occur in the real world are *irreversible*.

The irreversibility of almost all natural phenomena is a basic fact of experience. Just think of trying to unbreak an egg, grow younger, make a river flow uphill or unstir the milk from your tea. You simply cannot make these things go backwards. But this raises a curious paradox. If the underlying laws that govern the activity of each atom of these systems are reversible, what is the origin of the irreversibility?

The hint of an answer was found in the mid nineteenth century with the study of thermodynamics. Physicists interested in the performance of heat engines had formulated a number of laws related to the exchange of heat and its conversion to other forms of energy. Of these, the so-called *second law of thermodynamics* held the clue to the arrow of time. In its original form the second law states, roughly speaking, that heat cannot flow on its own from cold to hot bodies. This is, of course, very familiar in ordinary experience. When we put ice in warm water, the water melts the ice, because heat flows from the warm liquid into the cold ice. The reverse process, where heat flows out of the ice making the water even warmer, is never observed.

These ideas were made precise by defining a quantity called *entropy*, which can be thought of, very roughly, as a measure of the potency of heat energy. In a simple system such as a flask of water or air, if the temperature is uniform throughout the flask, nothing will happen. The system remains in an unchanging state called *thermodynamic equilibrium*. The flask will certainly contain heat energy, but this energy cannot do anything. It is impotent. By contrast, if the heat energy is concentrated in a 'hot spot' then things will happen, such as convection and changes in density. These events will continue until the heat dissipates and the system reaches equilibrium at a uniform temperature.

The definition of entropy for such a system involves both heat energy and temperature, and is such that the greater the 'potency' of the heat energy, the lower the entropy. A state of thermodynamic equilibrium, for which the heat energy is impotent, has maximum entropy. The second law of thermodynamics can then be expressed as follows: *In a closed system, entropy never decreases.* If a system starts out, for example, with a non-uniform temperature distribution, i.e. at relatively low entropy, heat will flow and the entropy will rise until it reaches a maximum, at which point the temperature will be uniform and thermodynamic equilibrium will be achieved.

The restriction to a closed system is an important one. If heat or other forms of energy can be exchanged between the system and its environment then the entropy can certainly be decreased. This is precisely what happens in a refrigerator, for example, where heat is extracted from cold bodies and delivered to the warm environment. There is, however, a price to be paid, which in the case of the refrigerator is the expenditure of energy. If this price is taken into account by including the refrigerator, its power supply, the surrounding atmosphere, etc. in one large system, then taking everything into account, the total entropy will rise, even though locally (within the refrigerator) it has fallen.

Chance

A very useful way of pinning down the operation of the second law is to study the exchange of heat between gases. In the nineteenth century the kinetic theory of gases was developed by James Clerk Maxwell in Britain and Ludwig Boltzmann in Austria. This theory treated a gas as a huge assemblage of molecules in ceaseless chaotic motion, continually colliding with each other and the walls of the container. The temperature of the gas was related to the level of agitation of the molecules and the pressure was attributed to the incessant bombardment of the container walls.

With this vivid picture, it is very easy to see why heat flows from hot to cold. Imagine that the gas is hotter in one part of the vessel than another. The more rapidly moving molecules in the hot region will soon communicate some of their excess energy to their slower neighbours through the repeated collisions. If the molecules move at random, then before long the excess energy will be shared out more or less evenly, and spread throughout the vessel until a common level of agitation (i.e. temperature) is reached.

The reason we regard this smoothing out of temperature as irreversible is best explained by analogy with card shuffling. The effect of the molecular collisions is akin to the random rearrangement of a deck of cards. If you start out with cards in a particular order – for example, numerical and suit sequence – and then shuffle the deck, you would not expect that further shuffling would return the cards to the original orderly sequence. Random shuffling tends to produce a jumble. It turns order into a jumble, and a jumble into a jumble, but practically never turns a jumble into order.

One might conclude that the transition from an orderly card sequence to a jumble is an irreversible change, and defines an arrow of time: order→ disorder. This conclusion is, however, dependent on a subtlety. There is an assumption that we can recognize an ordered sequence when we see one, but that we do not distinguish one jumbled sequence from another. Given this assumption it is clear that there will be very many more sequences of cards that are designated 'jumbled' than those designated 'ordered'. It then follows that so long as the shuffling is truly random, jumbled sequences will be produced much more often than ordered sequences – because there are so many more of them. Another way of expressing this is to say that a sequence picked at random is far more likely to be jumbled than ordered.

The card shuffling example serves to introduce two important ideas. First, the concept of irreversibility has been related to order and disorder, which are partly subjective concepts. If one regarded all card sequences as equally significant there would be no notion of 'vast numbers of jumbled states', and shuffling would be considered as simply transforming one particular card sequence into another particular sequence. Secondly, there is a fundamental *statistical* element involved. The transition from order to disorder is not *absolutely* inevitable; it is something which is merely *very probable* if the shuffling is random. Clearly, there is a tiny but non-zero chance that shuffling a jumbled card sequence will transform it into suit order. Indeed, if one were to shuffle long enough, *every* possible sequence would eventually crop up, including the original one.

It seems, then, that an inexhaustible shuffler would eventually be able to get back to the original ordered sequence. Evidently the destruction of the orderly initial state is not irreversible after all: there is nothing intrinsically time-asymmetric about card shuffling.

Is the arrow of time therefore an illusion here? Not really. We can certainly say that *if* the cards are initially ordered and then shuffled a few times, it is overwhelmingly likely that the deck will be less ordered afterwards than before. But the arrow clearly does not come from the shuffling as such; rather, it owes its origin to the special, orderly nature of the initial state.

These ideas carry over in a fairly straightforward way to the case of a gas. The state of the gas at any instant is given by specifying the position and velocity of every molecule. If we could really observe a gas at the molecular level, and if we regarded each state as equally significant, there would be no arrow of time. The gas would merely 'shuffle itself' from one particular state to another. However, in practice we are not interested in the exact position and velocity of every molecule, nor could we actually observe them. Most states we regard as simply 'jumbled', and do not distinguish between them. If the gas is in a relatively ordered state initially (such as the state which is hot at one end and cold at the other), then it is overwhelmingly probable that the molecular collisions will produce a less orderly state, for the simple reason that there are so many more ways for the gas to be jumbled than ordered.

It is possible to quantify all this by computing the number of ways the molecules can be arranged at the microscopic level without our noticing any change at the macroscopic level. This is a subject called *statistical mechanics*. One divides up the volume of the box into little cells representing the limit of resolution of our instruments. A molecule is then considered to be either in a particular cell or not. We do not worry about precisely where in the cell it is located. Something similar is done for the velocities. It is then a straightforward matter to work out the various permutations of molecules among cells. A state of the gas will now, from the point of view of the macroscopic observer, be given by specifying something like the numbers of molecules in each cell.

Some states of the gas will then be achievable in very few ways; for example, the state in which all the molecules are in one cell. Others will be achievable in a great many different ways. Generally, the less orderly the state is the greater the number of ways that the molecules may be distributed among the cells to achieve it.

One state will represent 'maximum disorder'. This is the state that can be achieved in the greatest number of ways. It then follows that if the states are 'shuffled' at random, the most probable state to result is the maximally disordered one. Once the gas has reached this state it is most probably going to remain in it, because further random shuffling is still more likely to reproduce this many-ways-to-achieve state than one of the rarer variety. The state of maximum disorder therefore corresponds to the condition of thermodynamic equilibrium.

A statistical quantity can be defined which represents the 'degree of disorder' of the gas. Boltzmann proved that so long as the molecular collisions are chaotic (in a rather precise sense) then this quantity would, with overwhelming probability, increase. Now this is precisely the same behaviour as the thermodynamic quality called entropy. Boltzmann had thus found a quantity in statistical mechanics that corresponds to the key thermodynamic quantity of entropy. His proof was thus a demonstration, at least in a simple model of a gas, of how the second law of thermodynamics goes about its business of driving up the entropy until it reaches a maximum.

The work of Maxwell and Boltzmann uncovered an arrow of time by introducing the concept of *chance* into physics. The French biologist Jacques Monod has described nature as an interplay of chance and necessity. The world of Newtonian necessity has no arrow of time. Boltzmann found an arrow hidden in nature's game of molecular roulette.

Is the universe dying?

Probably the most fearsome result ever produced in the history of science was first announced by the German physicist Hermann von Helmholtz in 1854. The universe, claimed Helmholtz, is doomed.

This apocalyptic prediction was based on the second law of thermodynamics. The remorseless rise in entropy that accompanies any natural process could only lead in the end, said Helmholtz, to the cessation of all interesting activity throughout the universe, as the entire cosmos slides irreversibly into a state of thermodynamic equilibrium. Every day the universe depletes its stock of available, potent energy, dissipating it into useless waste heat. This inexorable squandering of a finite and irretrievable resource implies that the universe is slowly but surely dying, choking in its own entropy.

We can witness the incessant advance of this cosmic decay in the way that the Sun and stars are burning up their reserves of nuclear fuel, pouring the energy released away into the depths of space. Sooner or later the fuel will run out and the stars will dim, leaving a cold, dark, lifeless universe. No new process, no mechanism, however ingenious, can alter this fate, because every physical process is subject to the imperative of the second law.

This gloomy prognosis is known as the 'heat death' of the universe, and it has strongly influenced science and philosophy over the last century. Consider, for example, the reaction of Bertrand Russell:[3]

> that all the labours of the ages, all the devotion, all the inspiration, all the noonday brightness of human genius, are destined to extinction in the vast death of the solar system, and the whole temple of Man's achievements must inevitably be buried beneath the debris of a universe in ruins – all these things, if not quite beyond dispute, are yet so nearly certain that no philosophy which rejects them can hope to stand. Only within the scaffolding of these truths, only on the firm foundation of unyielding despair, can the soul's habitation henceforth be safely built.

Some thinkers have balked at the ghastliness of the heat death, and sought an escape. The Marxist philosopher Friedrich Engels believed that in the end the second law of thermodynamics could be evaded:[4]

> In some way, which it will later be the task of scientific research to demonstrate, the heat radiated into space must be able to become transformed into another

form of motion, in which it can once more be stored up and rendered active. Thereby the chief difficulty in the way of the reconversion of extinct suns into incandescent vapour disappears.

Most scientists, however, have only confirmed the absolutely fundamental nature of the second law, and the hopelessness of avoiding the relentless rise of entropy. Sir Arthur Eddington put it thus:[5]

> The law that entropy always increases – the Second Law of Thermodynamics – holds, I think, the supreme position among the laws of Nature. If someone points out to you that your pet theory of the universe is in disagreement with Maxwell's equations – then so much the worse for Maxwell's equations. If it is found to be contradicted by observation – well, these experimentalists do bungle things sometimes. But if your theory is found to be against the Second Law of Thermodynamics I can give you no hope; there is nothing for it but to collapse in deepest humiliation.

It seems, then, that Boltzmann and his colleagues discovered an arrow of time, but one that points 'the wrong way' for many people's liking, in the direction of degeneration and death.

There exists alongside the entropy arrow another arrow of time, equally fundamental and no less subtle in nature. Its origin lies shrouded in mystery, but its presence is undeniable. I refer to the fact that the universe is *progressing* – through the steady growth of structure, organization and complexity – to ever more developed and elaborate states of matter and energy. This unidirectional advance we might call the optimistic arrow, as opposed to the pessimistic arrow of the second law.

There has been a tendency for scientists to simply deny the existence of the optimistic arrow. One wonders why. Perhaps it is because our understanding of complexity is still rudimentary, whereas the second law is firmly established. Partly also, perhaps it is because it smacks of anthropocentric sentimentality and has been espoused by many religious thinkers. Yet the progressive nature of the universe is an objective fact, and it somehow has to be reconciled with the second law, which is almost certainly inescapable. It is only in recent years that advances in the study of complexity, self-organization and cooperative phenomena has revealed how the two arrows can indeed co-exist.

3
Complexity

The modelling problem

'The universe is not made, but is being made continually. It is growing, perhaps indefinitely...' Thus wrote Henri Bergson,[1] one of the foremost philosophers of this century. Bergson recognized that new forms and structures are coming into being all the time, so that the universe is advancing, or evolving with a definite arrow of time. Modern science affirms this: the universe began in featureless simplicity, and grows ever more elaborate with time.

Although this unidirectional trend is apparent, it is not easy to identify the quality that is advancing. One candidate is complexity. The primeval universe was probably in a state of extreme – perhaps maximal – simplicity. At the present epoch, by contrast, complexity abounds on all scales of size from molecules to galactic superclusters. So there exists something like a law of increasing complexity. But the study of complexity is still very much in its infancy. The hope is that by studying complex systems in many different disciplines, new universal principles will be discovered that might cast light on the way that complexity grows with time.

When I was a child few people possessed central heating. One of the delights of rising from bed on a cold winter's day was to see the intricate tracery of ice patterns that adorned the bedroom window, sparkling in the morning sunlight. Even those who have not shared this experience will have marvelled at the elaborate structure of a snowflake with its striking combination of complexity and hexagonal symmetry.

The natural world abounds with complex structures that amalgamate regularity and irregularity: coastlines, forests, mountain chains, ice sheets, star clusters. Matter is manifested in a seemingly limitless variety of forms. How does one go about studying them scientifically?

A fundamental difficulty is that, by their very nature, complex forms have a high degree of individuality. We recognize a snowflake as a snowflake, but no two of them are the same. Conventional science attempts to explain things exactly, in terms of general principles. Any sort of explanation for the shape of a snowflake or a coastline could not be of this sort.

The Newtonian paradigm, which is rooted in that branch of mathematics – the differential calculus – that treats change as smooth and continuous, is not well adapted to deal with irregular things. The traditional approach to complicated, irregular systems is to model them by approximation to regular systems. The more irregular the real system is, the less satisfactory this modelling becomes. For example, galaxies are not distributed smoothly throughout space, but associate in clusters, strings, sheets and other forms that are often tangled and irregular in form. Attempts to model these features using Newtonian methods involve enormous computer simulations that take many hours even on modern machines.

When it comes to very highly organized systems, such as a living cell, the task of modelling by approximation to simple, continuous and smoothly varying quantities is hopeless. It is for this reason that attempts by sociologists and economists to imitate physicists and describe their subject matter by simple mathematical equations is rarely convincing.

Generally speaking, complex systems fail to meet the requirements of traditional modelling in four ways. The first concerns their formation. Complexity often appears abruptly rather than by slow and continuous evolution. We shall meet many examples of this. Secondly, complex systems often (though not always) have a very large number of components (degrees of freedom). Thirdly, they are rarely closed systems; indeed, it is usually their very openness to a complex environment that drives them. Finally, such systems are predominantly 'non-linear', an important concept that we shall look at carefully in the next section.

There is a tendency to think of complexity in nature as a sort of annoying aberration which holds up the progress of science. Only very recently has an entirely new perspective emerged, according to which complexity and irregularity are seen as the norm and smooth curves the exception. In the traditional approach one regards complex systems as complicated collections of simple systems. That is, complex or irregular systems are in principle *analysable* into their simple constituents, and the behaviour of the whole is believed to be reducible to the behaviour of the constituent parts. The new approach treats complex or irregular systems as primary in their own right. They simply cannot be 'chopped up' into lots of simple bits and still retain their distinctive qualities.

We might call this new approach synthetic or holistic, as opposed to analytic or reductionist, because it treats systems as wholes. Just as there are idealized simple systems (e.g. elementary particles) to use as building blocks in the reductionist approach, so one must also search for idealized complex or irregular systems to use in the holistic approach. Real systems can then be regarded as approximations to these idealized complex or irregular systems.

The new paradigm amounts to turning three hundred years of entrenched philosophy on its head. To use the words of physicist Predrag Cvitanović:[2] 'Junk your old equations and look for guidance in clouds' repeating patterns.' It is, in short, nothing less than a brand new start in the description of nature.

Linear and non-linear systems

Whatever the shortcomings of conventional modelling, a wide range of physical systems can, in fact, be satisfactorily approximated as regular and continuous. This can be often traced to a crucial property known as *linearity*.

A linear system is one in which cause and effect are related in a proportionate fashion. As a simple example consider stretching a string of

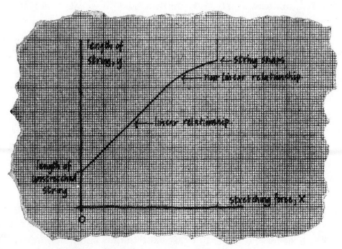

Figure 1. The length of an elastic string, y, is said to be 'linearly' related to the stretching force, x, when the graph of y against x is a straight line. For a real string non-linear behaviour sets in when the stretching becomes large.

elastic. If the elastic stretches by a certain length for a certain pull, it stretches by twice that length for twice the pull. This is called a linear relationship because if a graph is plotted showing the length of the string against the pulling force it will be a straight line (Figure 1). The line can be described by the equation $y = ax + b$, where y is the length of the string, x is the force, and a and b are constants.

If the string is stretched a great deal, its elasticity will start to fail and the proportionality between force and stretch will be lost. The graph deviates from a straight line as the string stiffens; the system is now *non-linear*. Eventually the string snaps, a highly non-linear response to the applied force.

A great many physical systems are described by quantities that are linearly related. An important example is wave motion. A particular shape of wave is described by the solution of some equation (mathematically this would be a so-called differential equation, which is typical of nearly all dynamical systems). The equation will possess other solutions too; these will correspond to waves of different shapes. The property of linearity concerns what happens when we superimpose two or more waves. In a linear system one simply adds together the amplitudes of the individual waves.

Most waves encountered in physics are linear to a good approximation, at least as long as their amplitudes remain small. In the case of sound waves, musical instruments depend for their harmonious quality on the linearity of vibrations in air, on strings, etc. Electromagnetic waves such as light and radio waves are also linear, a fact of great importance in telecommunications. Oscillating currents in electric circuits are often linear too, and most electronic equipment is designed to operate linearly. Non-linearities that sometimes occur in faulty equipment can cause distortions in the output.

A major discovery about linear systems was made by the French mathematician and physicist Jean Fourier. He proved that any periodic mathematical function can be represented by a (generally infinite) series of pure sine waves, whose frequencies are exact multiples of each other. This means that any periodic signal, however complicated, can be *analysed* into a sequence of simple sine waves. In essence, linearity means that wave motion, or any periodic activity, can be taken to bits and put together again without distortion.

Linearity is not a property of waves alone; it is also possessed by electric and magnetic fields, weak gravitational fields, stresses and strains in many materials, heat flow, diffusion of gases and liquids and much more. The greater part of modern science and technology stems directly from the fortunate fact that so much of what is of interest and importance in modern society involves linear systems. Roughly speaking, a linear system is one in which the whole is simply the sum of its parts. Thus, however complex a linear system may be it can always be understood as merely the conjunction or superposition or peaceful coexistence of many simple elements that are present together but do not 'get in each other's way'. Such systems can therefore be decomposed or analysed or reduced to their independent component parts. It is not surprising that the major burden of scientific research so far has been towards the development of techniques for studying and controlling linear systems. By contrast, non-linear systems have been largely neglected. In a non-linear system the whole is much more than the sum of its parts, and it cannot be reduced or analysed in terms of simple subunits acting together. The resulting properties can often be unexpected, complicated and mathematically intractable.

In recent years, though, more and more effort has been devoted to studying non-linear systems. An important result to come out of these investigations is that even very simple non-linear systems can display a remarkably rich and subtle diversity of behaviour. It might be supposed that complex behaviour requires a complex system, with many degrees of freedom, but this is not so. We shall look at an extremely simple non-linear system and find that its behaviour is actually infinitely complex.

Instant complexity

The simplest conceivable motion is that of a single point particle which jumps about abruptly from one location to another along a line. We shall consider an example of this where the motion is deterministic, that is, where each location of the point is completely determined by its previous location. It is then determined for all time, once the initial location is given, by specifying a procedure, or algorithm, for computing successive jumps.

To model the jumping motion mathematically one can label points on the line by numbers (*see* Figure 2) and then use a simple algorithm to

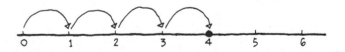

Figure 2. Each point on the line corresponds to a number. The 'particle' is a mobile point that moves along the line in hops, following an itinerary prescribed by an arithmetic algorithm. Here the algorithm is simply 'add one'.

generate a sequence of numbers. This sequence is then taken to correspond to successive positions of the particle, with each application of the algorithm representing one unit of time (i.e. 'tick of a clock'). To take an elementary example, if we start with the point at 0, and adopt the simple algorithm 'add one', we obtain the sequence 1,2,3,4,5,6,7, . . . which describes the particle jumping in equal steps to the right. This is an example of a linear algorithm, and the resulting motion is anything but complex.

At first sight it seems that to generate a complicated sequence of numbers requires a complicated algorithm. Nothing could be farther from the truth. Consider the algorithm 'multiply by two', which might yield the sequence 1,2,4,8,16, . . . As it stands this algorithm is also linear, and of limited interest, but a small alteration alters things dramatically.

Instead of ordinary doubling we shall consider 'clock doubling'. This is what you do when you double durations of time as told on a clock. The numbers on the clock face go from 1 to 12, then they repeat: 12 is treated as 0, and you start counting round again. If something takes 5 hours, starting at midday, it finishes at 5 o'clock. If it takes twice as long it finishes at 10 o'clock. Twice as long again takes us, not to 20, but round to 8 o'clock, because we start again from 0 when the hour hand crosses the 12.

What is happening here is that, rather than doubling a length, we are doubling an angle. When angles reach 360° we start back at 0. In terms of line intervals, it is equivalent to replacing an infinite line by a circle.

We are going to use clock doubling as an algorithm for generating the itinerary of a point that jumps on a line. The numbers on the 'clock', however, will be those that lie between 0 and 1 (*see* Figure 3). On reaching 1 we start back at 0 again. Doubling a number less than ½ proceeds as

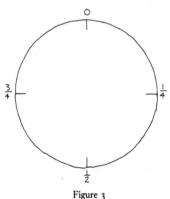

Figure 3

usual: for example 0.4 doubles to 0.8. But numbers greater than ½, when doubled, exceed 1, so we drop the 1 and retain only the decimal part. Thus 0.8 doubles to 1.6, which becomes 0.6. Although conventional doubling is linear, clock doubling has the crucial property of non-linearity.

The procedure of clock doubling is illustrated pictorially in Figure 4. Starting with the line segment 0 to 1, first stretch it to twice the length (Figure 4(a)). This corresponds to doubling the number. Now cut the stretched segment centrally (Figure 4(b)) and place the two halves exactly

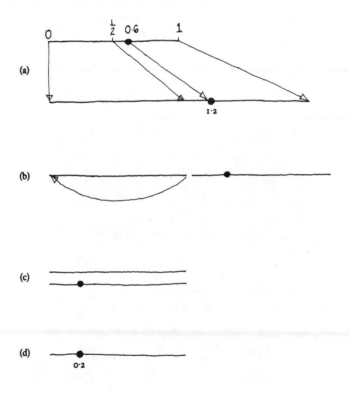

Figure 4 (a) The line interval 0 to 1 is stretched to twice its length: each number in the interval is doubled. (b) The stretched line is cut in the middle. (c) The two segments are stacked. (d) The stacked line segments are merged, thereby recovering an interval of unit length again. This sequence of operations is equivalent to doubling numbers and extracting only the decimal part. Shown as an example is the case of 0.6, which becomes 0.2.

on top of one another (Figure 4(c)). Finally merge the two segments into one to produce a line of the same length as you started with (Figure 4(d)). The whole operation can now be repeated for the next step of the algorithm. The procedure of successive squashing out and merging can be compared to rolling pastry.

To compute the detailed itinerary of a 'particle' under this algorithm you can either use a calculator, or use a diagram of the sort shown in Figure 5. The horizontal axis contains the line interval 0 to 1, and we start by selecting a point marked x_0. To generate the next point, x_1, go vertically from x_0 as far as the thick line, then horizontally to the broken line. Now read off the new value, x_1, on the horizontal axis. The procedure may now be repeated to find the next point, x_2, and so on.

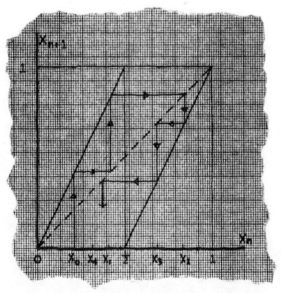

Figure 5. Predestiny versus prediction. The path shown generates in a completely deterministic way a sequence of numbers $x_1, x_2, x_3, \ldots$ The numbers can be envisaged as an itinerary for a particle that jumps about between 0 and 1 on the horizontal line. In spite of the fact that its itinerary for ever more is uniquely determined by the initial position x_0, for almost all choices of x_0 the particle moves randomly; its career inevitably becomes unpredictable unless x_0 is known *exactly* – which is impossible.

In spite of the simplicity of this algorithm it generates behaviour which is so rich, complex and erratic that it turns out to be completely unpredictable. In fact, in most cases the particle jumps back and forth in an apparently *random* fashion!

To demonstrate this it is convenient to make use of binary numbers. The binary system is a way of expressing all numbers using only two symbols, 0 and 1. Thus a typical binary number between 0 and 1 is 0.100101101000111101. There is no need for the reader to worry about how to convert ordinary base ten numbers into binary form. Only one rule will be needed. When ordinary numbers are multiplied by 10 one only needs to shift the decimal point one place to the right; thus $0.3475 \times 10 = 3.475$. Binary numbers have a similar rule except that it is multiplication by 2 rather than 10 that shifts the point. So 0.1011, when doubled, becomes 1.011. The rule adapts naturally to the doubling algorithm: successive applications applied to the number 0.1001011, for example, yield 0.001 011, 0.010 11, 0.1011, 0.011, 0.11 and so on (remembering to drop the 1 before the point if it appears).

If the interval 0 to 1 is represented by a line (*see* Figure 6) then numbers less than ½ lie to the left of centre, while numbers greater than ½ lie to the right. In binary, these correspond to numbers for which the first entry after the point is 0 or 1 respectively. Thus 0.1011 lies on the right and 0.01011 lies on the left. We could envisage two cells, or bins, labelled L and R for left and right intervals, and assign each number to either L or R depending on whether its binary expansion begins with 0 or 1. The doubling algorithm causes the particle to jump back and forth between L and R.

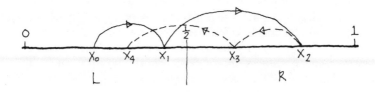

Figure 6. The line interval from 0 to 1 is here divided into two segments, L and R. As the particle jumps about along the line it may hop from L to R or vice versa. The complete LR sequence is exactly equivalent to the binary expansion of the initial number x_0. The sequence shown is that of the example depicted in Fig. 5.

Suppose we start with the number 0.011010001, which corresponds to a point in the left hand cell because the first entry after the decimal point is 0. The particle therefore starts out in L. When doubled, this number becomes 0.11010001, which is on the right, i.e. the particle jumps into R. Doubling again gives 1.1010001, and our algorithm requires that we drop the 1 before the decimal. The first entry *after* the decimal is 1, so the particle stays in R. Continuing this way we generate the jump sequence LRRLRLLLR.

It will be clear from the foregoing that the fate of the particle (i.e. whether it is in L or R) on the *n*th step will depend on whether the *n*th digit is a 0 or 1. Thus two numbers which are identical up to the *n*th decimal place, but differ in the *n*+1 entry, will generate the same sequence of L to R jumps for *n* steps, but will then assign the particle to different bins on the next step. In other words, two starting numbers that are very close together, corresponding to two points on the line that are very close together, will give rise to sequences of hops that eventually differ greatly.

It is now possible to see why the motion of the particle is unpredictable. Unless the initial position of the particle is known *exactly* then the uncertainty will grow and grow until we eventually lose all ability to forecast. If, for instance, we know the initial position of the particle to an accuracy of 20 binary decimal places, we will not be able to forecast whether it will be on the left or right of the line interval after 20 jumps. Because a *precise* specification of the initial position requires an *infinite* decimal expansion, *any* error will sooner or later lead to a deviation between prediction and fact.

The effect of repeated doublings is to stretch the range of uncertainty with each step (it actually grows exponentially), so that no matter how small it is initially it will eventually encompass the entire line interval, at which point all predictive power is lost. Thus the career of the point, although completely deterministic, is so exquisitely sensitive to the initial condition that any uncertainty in our knowledge, however small, suffices to wreck predictability after only a finite number of jumps. There is thus a sense in which the behaviour of the particle displays *infinite complexity*. To describe the career of the particle exactly would require specifying an infinite digit string, which contains an infinite quantity of information. And of course in practice one could never achieve this.

Although this simple example has the appearance of a highly idealized mathematical game, it has literally cosmic significance. It is often supposed that unpredictability and indeterminism go hand in hand, but now it can be seen that this is not necessarily so. One can envisage a completely deterministic universe in which the future is, nevertheless, unknown and unknowable. The implication is profound: even if the laws of physics are strictly deterministic, there is still room for the universe to be creative, and bring forth unforeseeable novelty.

A gambler's charter

A deep paradox lies at the heart of classical physics. On the one hand the laws of physics are deterministic. On the other hand we are surrounded by processes that are apparently random. Every casino manager depends on the 'laws of chance' to remain in business. But how is it possible for a physical process, such as the toss of a die, to comply with both the deterministic laws of physics *and* the laws of chance?

In the previous chapter we saw how Maxwell and Boltzmann introduced the concept of chance into physics by treating the motions of large assemblages of molecules using statistical mechanics. An essential element in that programme was the assumption that molecular collisions occur at random. The randomness in the motion of gas molecules has its origin in their vast numbers, which precludes even the remotest hope of keeping track of which molecules are moving where. Similarly in the throw of a die, nobody can know the precise conditions of the flip, and all the forces that act on the die. In other words, randomness can be attributed to the action of forces (or variables of some sort) that are in practice hidden from us, but which in principle are deterministic. Thus a Laplacian deity who could follow every twist and turn of a collection of gas molecules would not perceive the world as random. But for us mere mortals, with our limited faculties, randomness is inescapable.

The puzzle is, if randomness is a product of ignorance, it assumes a subjective nature. How can something subjective lead to *laws* of chance that legislate the activities of material objects like roulette wheels and dice with such dependability?

The search for the source of randomness in physical processes has been dramatically transformed by the discovery of examples such as the jumping particle. Here is a process which is unpredictable in true gambling fashion, yet makes no use of the notion of large numbers of particles or hidden forces. Indeed, one could hardly envisage a process more transparently simple and deterministic than that described in the previous section.

It is actually possible to prove that the activity of the jumping particle is every bit as random as tossing a coin. The argument given here follows the elegant discussion given by Joseph Ford of the Georgia Institute of Technology.[3] Ford's demonstration requires a short excursion into the theory of numbers. Returning to ordinary arithmetic for a moment, the interval from 0 to 1 obviously contains an infinite number of points which may be specified by an infinite collection of decimal numbers. Among these decimals are those of the fractions, such as $\frac{1}{2}$, $\frac{1}{3}$, $\frac{1}{5}$, etc. Some fractions possess finite decimal expansions, e.g. $\frac{1}{2} = 0.5$, while others, such as $\frac{1}{3}$, require an infinity of decimal places: $\frac{1}{3} = 0.333\ 333\ 333\ldots$ The finite strings can be regarded as simple cases of infinite strings by adding zeros: thus $\frac{1}{2} = 0.500\ 000 \ldots$ Note that all fractions either have finite decimal expansions followed by zeros, or else they eventually repeat periodically in some way: for example $\frac{3}{11} = 0.272\ 727\ 272 \ldots$ and $\frac{7}{13} = 0.538\ 461\ 538\ 461 \ldots$

Although every fraction has a decimal expansion, not all decimals can be expressed as fractions. That is, the set of all infinite decimals contains more numbers than the set of all fractions. In fact there are infinitely many more of these 'extra' decimals (known as 'irrational numbers') than there are fractions, in spite of the fact that there exist an infinity of fractions already. Some notable examples of irrational numbers are π, $\sqrt{2}$ and exponential e. There is no way that such numbers can be represented by fractions, however complicated.

Attempts to write out the number π as a digit string $(3.141\ 59 \ldots)$ always involve a certain degree of approximation as the string has to be truncated at some point. If a computer is used to generate ever more decimal places of π it is found that no sequence ever repeats itself periodically (in contrast with the decimal expansion of a fraction). Although this can be directly checked to only a finite number of decimal places, it can be proved that no systematic periodicity can ever appear. In other words the decimal places of π form a completely erratic sequence.

Returning now to binary arithmetic, we may say that all the numbers between 0 and 1 can be expressed by infinite strings of ones and zeros (after the point). Conversely, *every* string of ones and zeros, in whatever combination we choose, corresponds to a point somewhere on the interval.

Now we reach the key point concerning randomness. Imagine a coin with 0 marked on one side and 1 on the other. Successive tosses of this coin will generate a digit sequence, e.g. 01001 1010110 ... If we had an infinite number of such coins we would generate *all* infinite digit sequences, and hence *all* numbers between 0 and 1. In other words, the numbers between 0 and 1 can be regarded as representing all possible outcomes of infinite sequences of coin tosses. But since we are prepared to accept that coin tossing is random, then the successive appearances of ones and zeros in any particular binary expansion is as random as coin tossing. Translating this into the motion of the jumping particle one may say that its hops between L and R are as random as the successive flips of a coin.

A study of the theory of numbers reveals another important feature about this process. Suppose we pick a finite digit string, say 101101. All binary numbers between 0 and 1 that start out with this particular string lie in a narrow interval of the line bounded by the numbers 101101000000 ... and 101101111111 ... If we choose a longer string, a narrower interval is circumscribed. The longer the string the narrower the interval. In the limit that the string becomes infinitely long the range shrinks to nothing and a single point (i.e. number) is specified.

Let us now return to the behaviour of the jumping particle. If the example digit string 101101 occurs somewhere in the binary expansion of its initial position then it must be the case that at some stage in its itinerary the particle will end up jumping into the above line interval. And a similar result holds, of course, for any finite digit string.

Now it can be proved that *every* finite digit string crops up somewhere in the infinite binary expansion of *every* irrational number (strictly, with some isolated exceptions). It follows that if the particle starts out at a point specified by any irrational number (and most points on the line interval are specified by irrational numbers), then sooner or later it must hop into the narrow region specified by any arbitrary digit string. Thus, the particle is assured to visit every interval of the line, however narrow, at some stage during its career.

One can go further. It turns out that any given string of digits not only crops up somewhere in the binary expansion of (almost) every irrational number, it does so infinitely many times. In terms of particle jumps, this means that when the particle hops out of a particular interval of the line, we know that eventually it will return – and do so again and again. As this remains true however small the region of interest, and as it applies to *any* such region anywhere on the line interval, it must be the case that the particle visits every part of the line again and again; there are no gaps. Technically this property is known as ergodicity, and it is the key assumption that has to be made in statistical mechanics to ensure truly random behaviour. There it is justified by appealing to the vast numbers of particles involved. Here, incredibly, it emerges automatically as a property of the motion of a *single* particle.

The claim that the motion of the particle is truly random can be strengthened with the help of a branch of mathematics known as algorithmic complexity theory. This provides a means of quantifying the complexity of infinite digit strings in terms of the amount of information necessary for a computing machine to generate them. Some numbers, even though they involve infinite binary expansions, can be specified by finite computer algorithms. Actually the number π belongs to this class, in spite of the apparently endless complexity of its decimal expansion. However, most numbers require *infinite* computer programming information for their generation, and can therefore be considered infinitely complex. It follows that most numbers are actually unspecifiable! They are completely unpredictable and completely incalculable. Their binary expansions are random in the most fundamental sense. Clearly, if the motion of a particle is described by such a number it too is truly random.

The toy example of the jumping particle serves the very useful purpose of clarifying the relationship between complexity, randomness, predictability and determinism. But is it relevant to the real world? Surprisingly, the answer is yes, as we shall see in the next chapter.

4
Chaos

Pharaoh's dream

And Pharaoh said to Joseph, 'I have dreamed a dream, and there is no one who can interpret it; and I have heard it said of you that when you hear a dream you can interpret it ... Behold, in my dream I was standing on the banks of the Nile; and seven cows, fat and sleek, came up out of the Nile and fed in the reed grass; and seven other cows came up after them, poor and very gaunt and thin, such as I had never seen before in all the land of Egypt.'¹

Joseph's interpretation of Pharaoh's dream is famous: Egypt would experience seven years of plenty followed by seven lean years of famine. It was a prediction that earned him the position of Pharaoh's Grand Vizier. But is the story credible?

A study of population trends among crop pests, fish, birds and other species with definite breeding seasons reveals a wide variety of change, ranging from rapid growth or extinction, through periodic cycles, to apparent random drift. The cause of this varied behaviour provides valuable insight into a form of complexity that has recently been recognized to have universal significance.

The simplest example of population change is unrestrained growth, such as that observed in a small colony of insects on a large remote island, or among fish in a big pond, or bacteria reproducing in a protective culture. Under these circumstances the number, N, of individuals will double in a fixed time – the average reproduction cycle time. This type of accelerating population increase is known as exponential growth. There is also the converse case of exponential decline, which can occur if the environment contains inadequate resources to sustain the whole population. Both cases are illustrated in Figure 7. Intermediate scenarios exist where the population grows or shrinks towards an

Figure 7

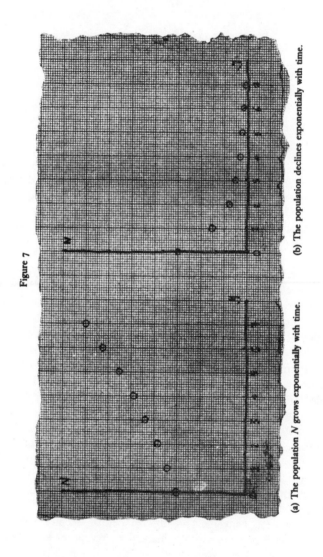

(a) The population N grows exponentially with time.

(b) The population declines exponentially with time.

optimum fixed value at which it stabilizes. Alternatively it may oscillate cyclically.

To see how oscillations arise, suppose in year one an island's insect population is small. There is plenty of food available for all, lots of breeding takes place and the population rises sharply. In year two the island is overrun with insects, and the limited food supply is over-subscribed. Result: a large death rate from starvation, followed by a low breeding rate. In year three the insect population is small again. And so on.

An interesting question is whether these, and other more complicated patterns of population change, can be modelled mathematically so that ecologists might be able to predict, as did Joseph, seven lean years. A simple approach is to suppose that the population each year is determined entirely by its size the year before, and then try a numerical experiment using certain fixed birth and death rates.

Imagine that the species has a certain breeding season once a year. Let us denote the population in year y by N_y. If breeding were unrestrained, the population in the following year, $y + 1$, would be proportional to that in year y, so we could write $N_{y+1} = aN_y$, where a is a constant depending on the reproductive efficiency of the species. The solution to this equation is readily obtained: it is the expected exponential growth.

In reality population growth is limited by food supply and other competitive factors, so we want to add a term to the above equation to allow for death, which will depress the breeding rate. A good approximation is to suppose that the probability of death for each individual is proportional to the total population, N_y. Thus the death rate for the population as a whole is proportional to N_y^2, say bN_y^2, where b is another constant. We are therefore led to study the equation $N_{y+1} = N_y(a - bN_y)$, which is known as the *logistic equation*.

The logistic equation can be regarded as a deterministic algorithm for the motion of a point on a line of the sort considered in Chapter 3. This is because if we pick a starting value N_0 for year 0 and use the right-hand side of the logistic equation to compute N_1, we can then put *this* value into the right-hand side to compute N_2, and so on. The string of numbers thereby obtained from this iteration form a deterministic sequence which can be envisaged as specifying the successive positions of a point on a line. The procedure is a simple job for a pocket calculator. The results, however, are far from simple.

To discuss them, it is first convenient to define $x = aN/b$ and study x instead of N. The equation becomes $x_{y+1} = ax_y(1 - x_y)$, and x is restricted to lie between the values 0 and 1, as in the example discussed in Chapter 3. One can draw a diagram similar to Figure 5, obtaining this time an inverted parabola in place of the pair of oblique lines.

If a is less than 1 the broken line lies entirely above the curve, as shown in Figure 8. To follow the fate of the population, pick a value x_0 to start, and go through the same procedure as described in connection with Figure 5; that is, go vertically to the curve, then horizontally to the broken line, and read off the following year's value, x_1. Then repeat for x_2, and so on. It should be clear from the diagram that, whatever the starting value, x_0, the population steadily declines and converges on zero. The resources of the island or pond are too meagre and extinction occurs.

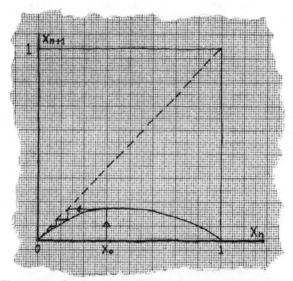

Figure 8. The sequence of numbers generated by the path shown converges on 0 for any choice of initial number X_0. This corresponds to a population inescapably destined for extinction due to inadequate resources. The year on year decline is similar to that shown in Figure 7 (b).

With the value of the parameter a chosen to be greater than 1, which corresponds to a somewhat larger island or pond with better resources, the broken line intersects the curve in two places (Figure 9). Following the same procedure as before, one now finds very different behaviour. In fact, as the value of a is varied, the solutions display a range of very complicated patterns of behaviour.

For *a* lying between 1 and 3 the population changes steadily until it stabilizes on the equilibrium value of $1 - 1/a$. A particular case is shown in Figure 9. See how the value of *x* gradually converges on the equilibrium value. The resulting population change is shown in Figure 10 (*a*).

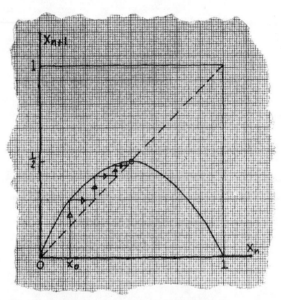

Figure 9. Choosing the value of the parameter *a* to be ½, the deterministic sequence converges on the fixed value = ½; it corresponds to a population that rises steadily and then stabilizes, as shown in Figure 10 (a).

For values of *a* greater than 3 (still more resources), the parabola is taller (Figure 11). A small initial population now begins by growing steadily, but then it begins to flip back and forth between two fixed values with a period of two years (see Figure 10 (*b*)). This is the Joseph effect. See how in Figure 11 the track of successive values converges on a box enclosing the intersection of the curve and oblique line.

As the island or pond is made larger, i.e. *a* is increased still more (above $1 + \sqrt{6} = 3.4495$ in fact), oscillations take place between *four* fixed values, with periodicity four years (Figure 10 (*c*)). For progressively higher values of *a* the period doubles again and again, more and more rapidly, until at a critical value, about 3.6, the population wanders about in a complex and highly erratic manner.

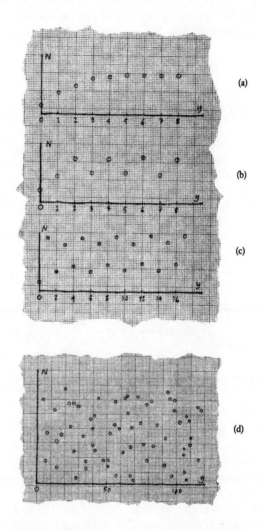

Figure 10. Possible population changes according to the logistic equation. (a) Steady rise to a stable equilibrium level. The year on year sequence can be generated from a diagram such as Fig. 9. (b) With a higher growth rate the population rises from its initially low value, then settles into a two-year oscillation – the Joseph effect. (c) With still higher growth rate, a four-year cycle occurs. (d) When the growth rate control parameter *a* has the value 4, the population changes chaotically, and is essentially unpredictable from year to year.

As one passes into the region beyond the critical value, x (hence N) displays very curious behaviour. It jumps in strict sequence between a number of bands of allowed values, but the precise positions visited within each band look entirely random. As a is increased further, the bands merge pairwise, so that the range of values over which N jumps erratically grows, until eventually a continuum is formed. As the value of a rises, this continuum spreads out. For the value $a = 4$, the continuum encompasses all values of x.

The situation at $a = 4$ is thus of particular interest. The changes in x look totally chaotic, i.e. the population seems to wander in a completely

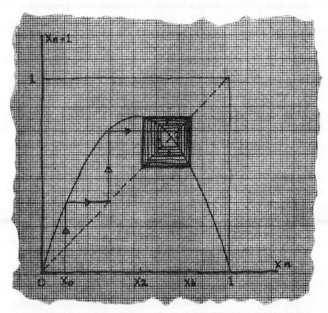

Figure 11. Choosing a to lie between 3 and $1 + \sqrt{6}$, the deterministic sequence rises, then converges on a 'limit cycle', represented by the bold square. The value of x thus settles down to alternate between the values x_a and x_b, corresponding to the oscillating population change shown in Fig. 10 (b).

random way (Figure 10 (*d*)). It is remarkable that such random behaviour can arise from a simple deterministic algorithm. It is also intriguing that certain bird and insect populations do indeed fluctuate from year to year in an apparently random fashion.

An interesting question is whether the complex behaviour for $a = 4$ is truly random or just very complicated. In fact it turns out to be truly random, as may readily be confirmed, because the equation can be solved exactly in this case. The change of variables $x_y = (1 - \cos 2\pi\theta_y)/2$ yields the simple solution that θ doubles every year. (That is $\theta_y = 2^y\theta_0$, where θ_0 is the starting value of θ.) It will be recalled from the discussion about 'clock doubling' given in Chapter 3 that successive doubling of an angle is equivalent to shifting binary digits one by one to the right, and that this implies truly random behaviour with infinite sensitivity on initial conditions.

This does not exhaust the extraordinarily rich variety of behaviour contained in the logistic equation. It turns out that the merging-band region between $a = 3.6$ and 4 is interrupted by short 'windows' of periodic or almost periodic behaviour. There is, for example, a narrow range (between 3.8284 and 3.8415) where the population displays a distinct three-year cyclic pattern. The reader is encouraged to explore this structure with a home computer.

Magic numbers

The kind of highly erratic and unpredictable behaviour being discussed here is known as *deterministic chaos*, and it has become the subject of intense research activity. It has been discovered that chaos arises in a wide range of dynamical systems, varying from heart beats to dripping taps to pulsating stars. But what has made chaos of great theoretical interest is a remarkable discovery by an American physicist, Mitchell Feigenbaum. Many systems approach chaotic behaviour through period doubling. In those cases the transition to chaos displays certain universal features, independent of the precise details of the system under investigation.

The features concerned refer to the rate at which chaotic behaviour is approached through the escalating cascade of period doublings discussed above. It is helpful to represent this pictorially by plotting x (or N) against a, as shown in Figure 12. For small a there is only one value of x that solves the equation, but at the critical point where $a = 3$ the solution curve suddenly breaks into two. This is called a *bifurcation* (sometimes a pitchfork bifurcation because of the shape), and it signals the onset of the first period doubling: x (or N) can now take two values, and it oscillates between them. Further on, more bifurcations occur, forming a 'bifurcation tree', indicating that x can wander over more and more values. The rate of bifurcations gets faster and faster, until at another critical value of a, an infinity of branches is reached. This is the onset of chaos.

The critical value at which chaotic behaviour starts is $3.5699\ldots$ As this point is approached the branchings get closer and closer together. If the gaps between successive branchings are compared one finds that each gap is slightly less than $1/4$ of the previous one. More precisely, the ratio tends to the fixed value $1/4.669\,201\ldots$ as the critical point is approached. Notice that this implies a 'self-similar' form, with a rate of convergence that is independent of scale, a fact that will turn out to be of some significance.

There is also a simple numerical relation governing the rate of shrinkage of the vertical gaps between the 'prongs of the pitchforks' on the bifurcation tree. Feigenbaum found that as the critical chaotic region is approached each gap is about $2/5$ of the previous one. (More precisely the ratio is $1/2.502\,9\ldots$)

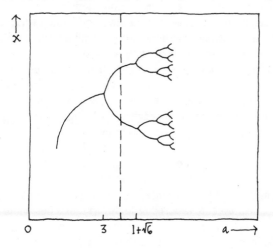

Figure 12. The road to chaos. Pick a value of a; draw a vertical. Where it cuts the 'bifurcation tree' gives the values of x (i.e. population) on which the population 'curve' settles down. The case shown gives two values, corresponding to a stable two-cycle of the sort shown in Fig. 10 (b). As a is increased, so the tree branches again and again, indicating an escalating cascade of period doubling. The converging multiplication of branchlets occurs in a mathematically precise fashion, dictated by Feigenbaum's numbers. Beyond the tracery of bifurcations lies chaos: the population changes erratically and unpredictably.

Feigenbaum came across the curious 'magic' numbers 4.669 201 . . . and 2.5029 . . . by accident, while toying with a small programmable calculator. The significance of these numbers lies not in their values but in the fact that they crop up again and again in completely different contexts. Evidently they represent a fundamental property of certain chaotic systems.

Driving a pendulum crazy

Random and unpredictable behaviour is by no means restricted to ecology. Many physical systems display apparently chaotic behaviour. A good example is provided by the so-called conical pendulum, which is an ordinary pendulum that is pivoted so as to be free to swing in any direction rather than just in a plane. The pendulum is the epitome of dynamical regularity – as regular as clockwork, so the saying goes. Yet it turns out that even a pendulum can behave chaotically. If it is driven by applying periodic forcing to the point of suspension, the bob (ball) is observed to undergo a remarkable range of interesting activity.

Before getting into this, a word should be said about why the system is non-linear. In the usual treatment of the pendulum, the amplitude of the oscillations are assumed to be small; the system is then approximately linear and its treatment is very simple. If the amplitude is allowed to become large, however, non-linear effects intrude. (Mathematically this is because the approximation $\sin \theta \sim \theta$ is breaking down.) Furthermore, frictional damping cannot be neglected if long-time behaviour is of interest, and indeed its effect is important here.

Although the pendulum is driven in one plane, the non-linearity can cause the bob to move in the perpendicular direction too, i.e. it is a system with two degrees of freedom. The bob thus traces out a path over a two-dimensional spherical surface. The principal feature of this system is that the bob will execute ordered or highly irregular behaviour according to the frequency of the driving force. A practical demonstration model has been made by my colleague David Tritton, who reports his observations as follows:[2]

The pendulum is started from rest with a driving frequency of 1.015 times the natural frequency. This initially generates a motion of the ball parallel to the point of suspension. This motion builds up in amplitude . . . until, after typically thirty seconds, the motion becomes two dimensional.

The path traced out by the bob ultimately settles down into a stable elliptical pattern, clockwise in some trials, anticlockwise in others. At this point the driving frequency is lowered to 0.985 times the natural frequency:

> The consequent change in the ball's motion is quickly apparent; its regularity is lost. Any few consecutive swings are sufficiently similar that one can say whether it is moving in a line, ellipse, or circle; but no such pattern is maintained for more than about five swings, and no particular sequence of changes is apparent. At any instant one might find the ball in linear, elliptical or circular motion with amplitude anywhere in a wide range; the line or major axis of the ellipse might have any orientation with respect to the driving motion. Any attempt to forecast what a look at the apparatus would reveal . . . would have little chance of success.

The foregoing example shows how a simple system can display very different patterns of behaviour depending on the value of a control parameter, in this case the driving frequency. A very slight alteration in the frequency can bring about a drastic transition from a simple, orderly and essentially predictable pattern of motion, to one that is apparently chaotic and unpredictable. We also found in the case of insect populations that the breeding rate a controlled whether the population grew steadily, oscillated, or drifted at random.

To investigate the matter in more detail it will be necessary to develop a helpful pictorial aid known as a phase diagram or portrait. This enables the general qualitative features of complex motion to be displayed in a simple diagrammatic form. As an example of the use of phase diagram we shall consider the simple pendulum. (This is a pendulum which swings in a plane, and must not be confused with the conical pendulum just described.)

A phase diagram consists of plotting a graph of the displacement of the bob, call it x, against the velocity of the bob, denoted v. At any instant of time, the state of the bob can be represented by a point on the phase diagram, specifying the position and velocity of the bob at that moment. Over a period of time the representative point traces out a curve. If frictional damping is neglected the curve consists of a simple closed loop (see Figure 13). Going once around the loop corresponds to one cycle of oscillation of the pendulum.

As the pendulum continues to swing it repeats its motion exactly, so the representative point just goes on round and round the loop, as indicated by the arrow. If friction is now introduced, the pendulum will steadily lose energy. As a result the amplitude of its oscillations will decay, and it will

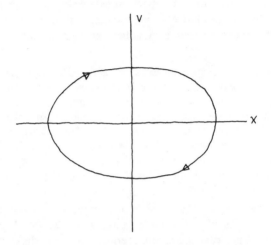

Figure 13. If the position *x* of the bob of a freely swinging pendulum is plotted against its velocity *v*, a curve known as the 'phase portrait' is traced out. In the absence of friction, the curve forms a closed loop (actually an ellipse).

eventually come to rest in the equilibrium position, i.e. with the bob vertically below the pivot. In this case the representative point spirals inwards, converging on a fixed point, known as an 'attractor', in the phase diagram (Figure 14).

Suppose now that the pendulum is driven periodically by some external force (but still restricting it to a plane – it is still a one degree of freedom problem). If the frequency of the driving force is different from the natural frequency of the pendulum the initial behaviour of the system will be rather complicated, because the driving force is trying to impose its motion on the pendulum's tendency to vibrate at its own natural frequency. The trajectory of the representative point will now be a complicated curve with a shape that depends on the precise details of the driving force.

However, because of the presence of frictional dissipation, the tussle between the two forms of motion will not last long. The efforts of the pendulum to assert its own motion become progressively damped, and the system settles down to be slavishly driven at the forcing frequency. The phase diagram therefore looks like Figure 15. The representative point, after executing some complex transient wiggles, winds itself progressively closer to the closed loop corresponding to the enslaved oscillations. And

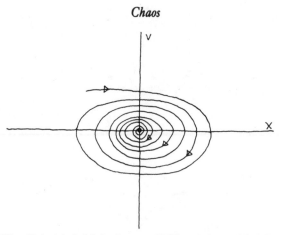

Figure 14. When friction is included, the phase portrait of the swinging pendulum changes to a spiral, converging on an 'attractor'. The spiral charts the decay of the pendulum's oscillations, as it dissipates energy through friction, eventually being damped to rest.

there it remains, going round and round, so long as the driving force continues. This closed loop is referred to as a *limit cycle*.

The final feature that we need is some non-linearity. Rather than allow the pendulum to vibrate out of the plane, we shall consider the simple expedient of making the restoring force on the pendulum non-linear (in fact, proportional to x^3). We need not worry about the nature of the agency that produces this non-linear force, but as we shall see its effect makes a crucial difference.

With a moderate amount of friction present, the behaviour of the pendulum is qualitatively similar to the previous case. The representative point starts out somewhere in the phase diagram, executes some complicated transient motion and then approaches a limit cycle. The main difference is that the limit cycle closed curve now has a couple of loops in it (Figure 16). Physically this is due to the driving force gaining temporary ascendancy over the restoring force and causing the pendulum to give a little backwards jerk each time it approaches the vertical.

Suppose now that the friction is progressively reduced. At a critical value of the damping parameter the phase diagram suddenly changes to the form shown in Figure 17. The limit cycle is still a closed loop, but it is now a 'double' loop, which means that the pendulum only repeats its motion exactly after two swings rather than one. In other words, the pendulum now executes a double swing, each swing slightly different, with a total period equal to twice the previous value. This phenomenon is

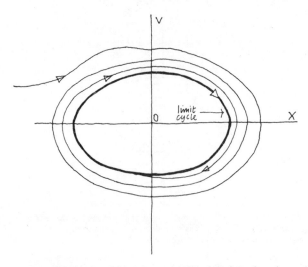

Figure 15. If the damped pendulum is driven by an external periodic force, then whatever its starting conditions, its phase path will wind round and round, eventually converging on a 'limit cycle' (bold line). When the limit cycle is reached, all memory of the starting conditions is lost and the pendulum's autonomy is completely subjugated by the external force.

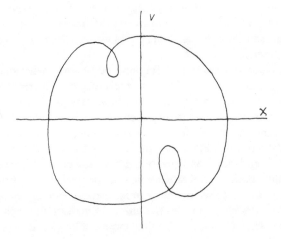

Figure 16. If a non-linearity is included in the driving force, the pendulum's motion becomes more complicated. The case shown is the limit cycle with a small cubic force added, which causes the pendulum to execute brief backward jerks, represented by the small loops.

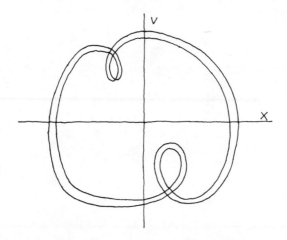

Figure 17. As the pendulum's damping is reduced beyond a critical value, period doubling suddenly occurs. The limit cycle now forms a closed double-loop.

referred to as 'period doubling', and it rings a bell. Exactly the same phenomenon was found in our study of insect populations.

With further reduction in the friction, a second abrupt period doubling occurs, so that the pendulum exactly repeats after *four* swings. As the friction is reduced further and further, so more and more period doublings take place (see Figure 18). Again, this is exactly what was found for the insect population problem.

The way in which the period doublings cascade together can be studied by taking a close-up look at a portion of the cycle in Figure 18. We can imagine looking through a little window in the phase diagram and seeing the representative point dart by, leaving a trace (Figure 19). After several transits the multiple-looped limit cycle would be complete and the pattern of lines would be redrawn. If a 'start line' is drawn across the 'window' we can keep track of where the phase trajectory intersects it each time around. Figure 19 (technically termed a Poincaré map after the French mathematician and physicist Henri Poincaré) shows a sequence of intersections. In the simplest case of high friction, there would only be one intersection, but with each period doubling the number will increase.

The positions of the intersections can be plotted against the value of

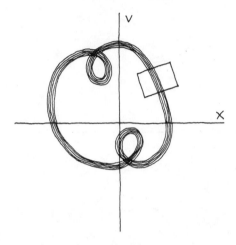

Figure 18. With further reduction in damping, the limit cycle splits and re-splits into a multi-loop, or band, indicating that the pendulum's motion is no longer discernably periodic, and predictability is breaking down. Its motion is approaching chaos.

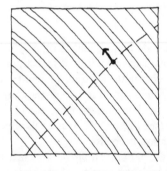

Figure 19. A magnified view of a section of the band shown in Fig. 18 shows the multiple tracks traced out by the representative point as it passes by again and again while executing one complete circuit of the limit cycle. The order of passage along the sequence of tracks is erratic. The 'start line' drawn across the tracks is analogous to the broken line in Fig. 12.

the declining friction, to show how the period doublings multiply as the damping gets less and less. One then obtains a 'bifurcation tree' diagram exactly like Figure 12. (Notice that the magnitude of the friction is plotted decreasing from left to right.) To the left of the figure there is only one intersection; this corresponds to the case depicted in Figure 16. At a critical value of the friction, the single line in Figure 12 suddenly *bifurcates*. This is the first period doubling, corresponding to Figure 17;

there are now two intersections. Further on, each branch bifurcates again, then again, with increasing rapidity. Eventually a value of the friction is reached at which the tree has sprouted an *infinite* number of branches. The motion of the pendulum is no longer periodic at all; it has to execute an infinity of different swings for the phase point to repeat its trajectory. The pendulum now moves in a highly disorderly and apparently random fashion. This is chaos once again.

We now recall that the onset of chaos in the logistic equation is described by the curious numbers 4.669201... and 2.5029... Although we are dealing here with a completely different system, nevertheless the same numbers crop up. This is not a coincidence. It seems that chaos has universal features, and that Feigenbaum's numbers are fundamental constants of nature. Thus although chaotic behaviour is, by definition, dauntingly difficult to model, there is still some underlying order in its manifestation, and we may obtain a broad understanding of the principles that govern this particular form of complexity.

Butterfly weather

Weather forecasters are the butt of many jokes. Although for most of us the weather is irrelevant to our daily lives, we nevertheless take a passionate interest in it, and tend to be derisory when the forecasters get it wrong. Indeed it is commonly believed (at least in Britain, where preoccupation with the weather – which is in any case rarely severe – is said to be a national obsession) that in spite of the vast computing power at their disposal, the meteorologists are more often wrong than right, or at least are no better than they were decades ago. (Which is not really true.) Indeed, many people have more faith in unorthodox methods, such as examining the condition of seaweed, or the habits of badgers or sparrows.

Although the weather seems very hard to predict, there is a widespread assumption that, seeing as the atmosphere obeys the laws of physics, an accurate mathematical model ought to be possible if only sufficient input data is available. But now this assumption is being challenged. It could be that the weather is intrinsically unpredictable in the long term.

The atmosphere behaves like a fluid heated from below, because the Sun's rays penetrate it on the way down and heat the Earth's surface, which then heats the overlying air by conduction and convection. Thus we find general vertical circulatory motion. Attempts to model atmospheric circulation mathematically go back a long way, but a landmark in this analysis occurred with the work of Edward Lorenz in 1963. Lorenz wrote down a system of equations that describe a simplified picture of atmospheric motion, and set about solving them.

What Lorenz found proved very disturbing for the forecasters. His equations, which have the crucial property of being non-linear, contain solutions that seem to be chaotic. It will be recalled from Chapter 3 that chaotic systems have the characteristic property of being essentially unpredictable. This is because solutions that start out very close together rapidly diverge, magnifying the domain of ignorance. Unless we know the initial state of the system to infinite precision, our predictability soon evaporates. This extreme sensitivity on the initial data implies that the circulatory patterns of the atmosphere might be ultimately decided by the most minute disturbance. It is a phenomenon sometimes called the butterfly effect, because the future pattern of weather might be decided by the mere flap of a butterfly's wings.

If Lorenz's equations capture a general property of atmospheric circulation, then the conclusion seems inescapable: long-term weather forecasting – be it by computer or seaweed divining – will never be possible, however much computing power may be deployed.

The unknowable future

Newton's clockwork universe – deterministic and mechanical – has always been hard to reconcile with the apparently random nature of many physical processes. As we have seen, Maxwell and Boltzmann introduced a statistical element into physics, but it has always been paradoxical how a theory based on Newtonian mechanics can produce chaos merely as the result of including large numbers of particles and making the subjective judgement that their behaviour cannot be observed by humans. The recent work on chaos provides a bridge between chance and necessity – between the probabilistic world of coin tossing and roulette and the clockwork universe of Newton and Laplace.

First, we have found that the existence of complex and intricate structures or behaviour does not necessarily require complicated fundamental principles. We have seen how very simple equations that can be handled on pocket calculators can generate solutions with an extraordinarily rich variety of complexity. Furthermore, quite ordinary systems in the real world (insect populations, pendula, the atmosphere) are found to closely conform to these equations and display the complexity associated with them. Secondly, it is becoming increasingly obvious that dynamical systems generally have regimes where their behaviour is chaotic. In fact, it seems that 'ordinary', i.e. non-chaotic, behaviour is very much the exception: *almost all* dynamical systems are susceptible to chaos. The evolution of such systems is exceedingly sensitive to the initial conditions, so that they behave in an essentially unpredictable and, for practical purposes, random fashion.

Although it is only comparatively recently that words such as 'scientific revolution' have been applied to the study of chaos, the essential discovery goes back to the turn of the century. In 1908, Henri Poincaré noted:[3]

A very small cause which escapes our notice determines a considerable effect that we cannot fail to see, and then we say that the effect is due to chance. If we knew exactly the laws of nature and the situation of the universe at the initial moment, we could predict exactly the situation of that same universe at a succeeding moment. But even if it were the case that the natural laws no longer had any secret for us, we could still only know the initial situation *approximately*. If that enabled us to predict the succeeding situation with the *same approximation*, that is all we require, and we should say that the phenomenon had been predicted, that is governed by the laws. But it is not always so; it may happen that small differences in the initial conditions produce very great ones in the final phenomena. A small error in the former will produce an enormous error in the latter. Prediction becomes impossible, and we have the fortuitous phenomenon.

It is important to emphasize that the behaviour of chaotic systems is not *intrinsically* indeterministic. Indeed, it can be proved mathematically that the initial conditions are sufficient to fix the entire future behaviour of the system exactly and uniquely. The problem comes when we try to specify those initial conditions. Obviously in practice we can never know *exactly* the state of a system at the outset. However refined our observations are there will always be *some* error involved. The issue concerns the effect this error has on our predictions. It is here that the crucial distinction between chaotic and ordinary dynamical evolution enters.

The classic example of Newtonian mechanistic science is the determination of planetary orbits. Astronomers can pinpoint the positions and velocities of planets only to a certain level of precision. When the equations of motion are solved (by integration) errors accumulate, so that the original prediction becomes less and less reliable over the years. This rarely matters, of course, because astronomers can keep updating the data and reworking their calculations. In other words, the calculations are always well ahead of the events. Eclipses of the Sun, for example, are reliably predicted for many centuries to come.

Typically the errors in these ordinary dynamical systems grow in proportion to time (i.e. linearly). By contrast, in a chaotic system the errors grow at an escalating rate; in fact, they grow exponentially with time. The randomness of chaotic motion is therefore *fundamental*, not merely the result of our ignorance. Gathering more information about the system will not eliminate it. Whereas in an ordinary system like the solar system the calculations keep well ahead of the action, in a chaotic system more and more information must be processed to maintain the same level of accuracy, and the calculation can barely keep pace with the actual events. In other words, all power of prediction is lost. The conclusion is that the system itself is its own fastest computer.

Joseph Ford likes to think of the distinction between ordinary and chaotic systems in terms of information processing. He points out that if we regard the initial conditions as 'input information' for a computer simulation of the future behaviour, then in an ordinary system we are rewarded for our efforts by having this input information converted into a very large quantity of output information, in the form of reasonably accurately predicted behaviour for quite a while ahead. For a chaotic system, however, simulation is pointless, because we only get the same amount of information out as we put in. More and more computing power is needed to tell us less and less. In other words we are not predicting anything, merely describing the system to a certain limited level of accuracy as it evolves in real time. To use Ford's analogy, in the computation of chaotic motion, our computers are reduced to xerox machines. We cannot determine a chaotic path unless we are first given that path.

To be specific, suppose a computer of a certain size takes an hour to compute a chaotic orbit of some particle in motion to a certain level of accuracy for one minute ahead. To compute to the same level of accuracy two minutes ahead might then require ten times the input data, and take ten hours to compute. For three minutes ahead one would then need 100 (i.e. 10^2) times as much data, and the calculation would take 100 hours; for four minutes it would take 1000 hours, and so on.

Although the word chaos implies something negative and destructive, there is a creative aspect to it too. The random element endows a chaotic system with a certain freedom to explore a vast range of behaviour patterns. Indeed, chaos can be employed in an efficient strategy for solving certain mathematical and physical problems. It is also seemingly used by nature herself, for example in solving the problem of how the body's immune system recognizes pathogens.

Furthermore, the occurrence of chaos frequently goes hand in hand with the spontaneous generation of spatial forms and structures. A beautiful example concerns the famous red spot on the surface of the planet Jupiter, a feature caused by swirling gases in the Jovian atmosphere. Computer simulations suggest that any particular element of fluid in the vicinity of the spot behaves chaotically and hence unpredictably, yet the gases as a whole arrange themselves into a stable coherent structure with a discrete identity and a degree of permanence. Another example, to be discussed further in Chapter 6, concerns the vortices and other features observed in the flow of a turbulent fluid.

These considerations show that nature can be *both* deterministic in principle, and random. In practice, however, determinism is a myth. This is a shattering conclusion. To quote Prigogine:[4]

> The basis of the vision of classical physics was the conviction that the future is determined by the present, and therefore a careful study of the present permits an unveiling of the future. At no time, however, was this more than a theoretical possibility. Yet in some sense this unlimited predictability was an essential element of the scientific picture of the physical world. We may perhaps even call it the founding myth of classical science. The situation is greatly changed today . . .

Joseph Ford makes the same point more picturesquely:[5]

> Unfortunately, non-chaotic systems are very nearly as scarce as hen's teeth, despite the fact that our physical understanding of nature is largely based upon their study . . . For centuries, randomness has been deemed a useful, but subservient citizen in a deterministic universe. Algorithmic complexity theory and nonlinear dynamics together establish the fact that determinism actually reigns only over a quite finite domain; outside this small haven of order lies a largely uncharted, vast wasteland of chaos where determinism has faded into an ephemeral memory of existence theorems and only randomness survives.

The conclusion must be that even if the universe behaves like a machine in the strict mathematical sense, it can still happen that genuinely new and in-principle unpredictable phenomena occur. If the universe were a linear Newtonian mechanical system, the future would, in a very real sense, be contained in the present, and nothing genuinely new could happen. But in reality our universe is not a linear Newtonian mechanical system; it is a chaotic system. If the laws of mechanics are the only organizing principles shaping matter and energy then its future is unknown and in principle unknowable. No finite intelligence, however powerful, could anticipate what new forms or systems may come to exist in the future. The universe is in some sense open; it cannot be known what new levels of variety or complexity may be in store.

5
Charting the Irregular

Fractals

'Clouds are not spheres, mountains are not cones.' Thus opens the book *The Fractal Geometry of Nature*, one of the most important recent contributions to understanding form and complexity in the physical universe. Its author is Benoit Mandelbrot, an IBM computer scientist who became fascinated with the challenge of describing the irregular, the fragmented and the complex in a systematic mathematical way.

Traditional geometry is concerned with regular forms: straight lines, smooth curves, shapes with perfect symmetry. At school we learn about squares, triangles, circles, ellipses. Yet nature rarely displays such simple structures. More often we encounter ragged edges, broken surfaces or tangled networks. Mandelbrot set out to construct a geometry of *irregularity* to complement the geometry of regularity that we learn at school. It is called fractal geometry.

A useful starting point in the study of fractals is the very practical problem of measuring the length of a coastline, or the frontier between two countries that includes sections of rivers. It is obvious that the length of coastline between, say, Plymouth and Portsmouth must be greater than the straight line distance between these two ports, because the coast wiggles about. Reference to an atlas would provide one estimate of the length of this irregular curve. If, however, one were to consult a more detailed Ordnance Survey map, a riot of little wiggles, too small to show up in the atlas, would be revealed. The coastline seems longer than we first thought. An inspection 'on the ground' would show yet more wiggles, on an even smaller scale, and the distance estimate would grow yet again. In fact, it soon becomes clear that the length of a coastline is a very ill-defined concept altogether, and could in a sense be regarded as infinite.

This simple fact causes endless confusion for geographers and governments, who often quote wildly different figures for the lengths of coastlines or common land frontiers, depending on the scale at which the distances are measured. The trouble is, if all coastlines have effectively infinite length, how could one compare the lengths of two different coastlines? Surely there must be some sense in which, say, the coast of America is longer than that of Britain?

One way of investigating this is to examine highly irregular curves that can be defined geometrically in a precise way. An important clue here is that if somebody shows you a map of a piece of unfamiliar coastline, it is usually impossible to deduce the scale. In fact, the degree of wiggliness seems generally to be independent of scale. A small portion of the coastline of Britain, for example, blown up in scale, looks more or less the same as does a larger section on a coarser scale. If small intervals of a curve are similar to the whole it is called self-similar, and it indicates a fundamental *scaling* property of the curve. (We have already met self-similarity in the way in which period doubling cascades into chaos.)

An explicit example of an irregular self-similar geometrical form was invented by the mathematician von Koch in 1904. It is constructed by an infinite sequence of identical steps, starting with an equilateral triangle (Figure 20). In the first step new equilateral triangles are erected

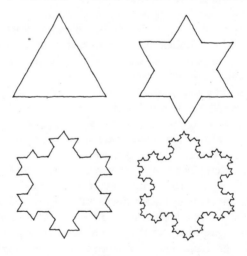

Figure 20. The Koch 'snowflake' is built by erecting successive triangles on the sides of larger triangles. In the limit of an infinite number of steps the perimeter becomes a fractal, with the weird property that it has a kink at every point.

symmetrically on the sides of the original, making a Star of David. Then the operation is repeated, leading to something reminiscent of a snowflake. The procedure is continued, *ad infinitum*. The end product is a continuous 'curve' containing an infinite number of infinitesimal kinks or excursions: a so-called Koch curve. It is almost impossible to visualize; it is a monstrosity. For example, it possesses no tangent, because the 'curve' changes direction abruptly at every point! It is therefore, in a sense, infinitely irregular. In fact, the Koch curve is so unlike the curves of traditional geometry that mathematicians initially recoiled in horror.

In the usual sense the length of the Koch curve is infinite; all those little triangular excursions sum without limit as the scale approaches zero. However, it possesses the important property of exact self-similarity. Magnify any portion of the Koch curve and it is completely identical to the whole; and this remains true however small the scale on which we examine it. It is this feature which enables us to get to grips with the concept of the length of a highly irregular curve.

Because the Koch curve is built in steps, we can keep track of precisely how the length of the curve grows with each step. Suppose the length of each side is l, then the total length around the curve at any given step can be obtained by multiplying l by the number of sides. The result is beautifully simple: l^{1-D}. Here the symbol D is shorthand for the number log 4/log 3, which is about 1.2618. Thus the length of the Koch curve is roughly $l^{-0.2618}$, which (because of the minus sign in the power) means that the length goes to infinity as l goes to zero.

The Koch curve has infinite length because its excursions and wiggles are so densely concentrated. It somehow 'visits' infinitely many more points than a smooth curve. Now a *surface* seems to have more points than a line because a surface is two dimensional whereas a line is only one dimensional. If we tried to cover a surface with a continuous line, zigzagging back and forth, we would certainly need to make it infinitely long because the line has zero thickness. (Actually the task is impossible.) With all those wiggles and kinks the Koch curve is somehow trying to be like a surface, although it doesn't quite make it because the perimeter certainly has zero area. This suggests that the Koch curve is best thought of as an object that somehow lies between being a line and a surface. It can, in fact, be described as having a dimension that lies *between* 1 and 2.

The idea of fractional dimensionality is not as crazy as it first seems. It was placed on a sound mathematical footing by F. Hausdorff in 1919. Its rigorous mathematical justification need not concern us. The point is that if one accepts Hausdorff's definition of dimensionality, then it permits certain mathematical objects (such as the Koch curve) to have fractional dimension, whereas 'normal' curves, surfaces and volumes still have the expected dimensions 1,2,3.

Using Hausdorff dimensionality provides a meaningful measure of the length of the Koch curve. The procedure is simple: the length of the curve is defined to be l^D times the number of segments of length l, where D is the Hausdorff dimension. For the Koch curve $D = \log 4/\log 3 = 1.2818\ldots$, and the curve may now be considered to have the finite length $l^D \times l^D = 1$, which is much more reasonable. Thus, using Hausdorff's definition of dimensionality, the Koch curve has dimension $1.2818\ldots$

Mandelbrot has coined the word *fractal* for forms like the Koch curve that have dimension (usually fractional) greater than naïvely expected. Mathematicians have catalogued a great many fractals, and Mandelbrot has generated many more. The question is, are they of interest only to mathematicians, or are there fractal structures in the real world? The Koch curve is only meant to be a crude model for a coastline, and further processing and refinement is necessary before realistic coastal shapes are generated. Nevertheless, it could be argued that the approximation of a coastline to a fractal is better than its approximation to a smooth curve, and so fractals provide a more natural starting point for the modelling of such forms.

Actually it was not Mandelbrot who first pointed out the formula l^{1-D} for coastlines. It was originally discovered by Lewis Fry Richardson, the eccentric uncle of the actor Sir Ralph Richardson. He was variously a meteorologist, physicist and psychologist with an interest in studying oddities off the beaten path. His study of coastlines uncovered the above-mentioned scaling law, and he was able to discover different values for the constant D for various familiar coastal regions, including Britain, Australia and South Africa.

Using computers to generate fractal curves and surfaces (the latter having dimension between 2 and 3) Mandelbrot has published beautiful pictures reminiscent of many familiar forms and structures. In his books and articles one finds islands, lakes, rivers, landscapes, trees, flowers, forests, snowflakes, star clusters, foam, dragons, veils and much else. His results are particularly striking when colour coded, and some abstract forms have considerable artistic appeal.

Fractals find many and diverse applications in physical science. They are especially significant in systems where statistical or random effects occur: for example, the famous Brownian motion, wherein a small particle suspended in a fluid zigzags about under the bombardment of its surfaces by the surrounding molecules. But fractals have also been applied to other subjects, such as biology and even economics.

The nearest thing to nothing known to man

A particularly intriguing fractal is known as the Cantor set, after the mathematician Georg Cantor, who also invented the subject called set theory. It is interesting to note in passing that Cantor's mathematical studies led him into such strange territory that there were serious medical reasons to doubt his sanity, and his work was denigrated by his contemporaries.

Like the Koch curve, Cantor's set is self-similar, and is built up in successive steps. The procedure is illustrated in Figure 21. Starting with a line of unit length, the middle one-third is cut out. Then the middle thirds from the remaining pieces are similarly removed, then their middle thirds, and so on, *ad infinitum*. (A subtlety is that the end points of the excised intervals, e.g. 1/3, 2/3, must be left behind.)

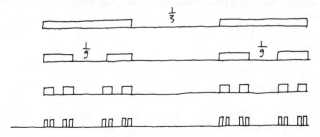

Figure 21. Constructing the Cantor 'dust' fractal. Shown are the first steps in an infinite sequence of excisions which turns the continuous line interval into a set containing gaps on every scale.

Now it might be supposed that this relentless robbing of segments would eventually deprive the line of all its parts, save possibly for isolated points. Certainly the end product has zero length, which seems to suggest that Cantor's set has dimension zero, this being the dimension of a collection of isolated points. Surprisingly such is not the case. It can be shown that Cantor's set is a fractal, with dimension log 2/log 3 = 0.6309 . . . In other words it is more than merely an infinite collection of unextended points, but it is not enough to achieve the actual extension of a continuous line – a source of much bafflement when its properties were first being explored.

By changing the fraction of line removed each time from $1/3$, it is possible to generate sets with dimension anywhere between 0 and 1. Those with dimension close to 1 are rather densely filled with points, whereas those with dimension near zero are relatively sparse.

For decades Cantor's set was dismissed as nothing more than a mathematical curiosity – or should one say monstrosity? Mandelbrot, however, argues that it corresponds to a good approximation to things in the real world. His interest in the set was first aroused by the study of intermittent noise in digital communication systems, where each burst of noise can be analysed as intermittent sub-bursts containing intermittent sub-sub-bursts, and so on, in a self-similar scaling fashion.

A more concrete example is provided by the ring system of Saturn. Although photographs make the rings look solid from afar, they are in fact composed of small particles rather sparsely distributed. Indeed, astronomers have no difficulty viewing stars *through* the rings. As early as 1675 the astronomer Giovanni Cassini discovered a gap in Saturn's rings, and over the years more gaps were discerned as the planet came to be studied in finer detail. Recently, American spacecraft have flown by Saturn and photographs have revealed thousands of finer and finer divisions. Rather that a continuous sheet, Saturn is in reality surrounded by a complex system of rings within rings – or gaps within gaps – reminiscent of Cantor's set.

The most complex thing known to man

The final fractal that we shall consider is named after Mandelbrot – the Mandelbrot set. It exists as a curve that forms the boundary of a region of a two-dimensional sheet called the complex plane, and has been described as the most complex object in mathematics. As so often in this subject the actual procedure for generating the Mandelbrot set is disarmingly simple. One merely keeps repeating an elementary mapping process. Points in the sheet that lie outside the region get mapped off to infinity, while points within cavort about in an incredibly intricate manner.

Points in a surface can be located by a pair of numbers, or coordinates (e.g. latitude and longitude). Let us denote these by x and y. The required mapping then merely consists of picking a fixed point in the surface, say x_0, y_0, and then replacing x by $x^2 - y^2 + x_0$ and y by $2yx + y_0$. That is, the point with coordinates x and y gets 'mapped' to the point with these new coordinates. (For readers familiar with complex numbers the procedure is simpler still: the mapping is from z to $z^2 + c$, where z is a general complex number, and c is the number $x_0 + iy_0$.) The Mandelbrot set can then be generated by starting with the coordinates $x = 0, y = 0$, and repeatedly applying the mapping, using the output coordinates from each go as the new x and y input for the next go. For most choices of x_0 and y_0 repeated mapping sends the point of interest off to infinity (and in particular, out of the picture). There exist choices, however, where this does not happen, and it is these points that form the Mandelbrot set.

To explore the structure of the Mandelbrot set a computer with colour graphics should be used. The forms that appear are breathtaking in their variety, complexity and beauty. One sees an astonishingly elaborate tracery of tendrils, flames, whorls and filigrees. As each feature is magnified and remagnified, more structure within structure appears, with new shapes erupting on every scale. The exceedingly simple mathematical prescription for generating the Mandelbrot set is evidently the source of an infinitely rich catalogue of forms.

Examples like the Mandelbrot set and the repeated mapping of points on a line discussed in Chapter 3 attest to the fact that simple procedures can be the source of almost limitless variety and complexity. It is tempting to believe that many of the complex forms and processes encountered in nature arise this way. The fact that the universe is full of complexity does not mean that the underlying laws are also complex.

Strange attractors

One of the most exciting scientific advances of recent years has been the discovery of a connection between chaos and fractals. Indeed, it is probably in the realm of chaotic systems that fractals will make their biggest scientific impact.

To understand the connection we have to go back to the discussion of the pendulum given in Chapter 4, and the use of phase diagrams as portraits of dynamical evolution. An important concept was that of the *attractor* – a region of the diagram to which the point representing the motion of the system is attracted. Examples were given of attractors that were points, or closed loops. It might be supposed that points and lines are the only possible sorts of attractor, but this is not so. There also exists the possibility of *fractal* attractors.

Fractal attractors are attracting sets of points in the phase diagram that have dimension lying between 0 and 1. When the representative point enters a fractal attractor it moves about in a very complicated and essentially random way, indicating that the system behaves chaotically and unpredictably. Thus, the existence of fractal attractors is a signal for chaos. For example, the first fractal attractor to be discovered was for Lorenz's system of equations mentioned at the end of Chapter 4.

In 1971 two French physicists, David Ruelle and F. Takens, seized upon these ideas and applied them to the age-old problem of turbulence in a fluid. In a pioneering paper they argued that the onset of turbulence can be explained as a result of a transition to chaotic behaviour, though via a route somewhat different from the period doubling discussed in Chapter 4. This bold assertion is quite at odds with the traditional understanding of turbulence. Clearly the relationship between chaos and turbulence will receive increasing attention.

Another example concerns the non-linear driven pendulum discussed in Chapter 4. This system is distinctive in the way it approaches chaos through an infinite cascade of period doublings. The path in the phase diagram winds round and round more and more often before closing. When chaos is reached, there are an infinite number of loops forming a finite band (see Figure 18). This is analogous to the problem discussed earlier of trying to fill out a two-dimensional surface with an infinity of widthless lines. In fact the band is a fractal (compare the rings of Saturn), and a section through it would be punctured by an infinite set of points forming a Cantor set.

When systems which possess fractal attractors were first studied, their peculiar properties seemed hard to comprehend, and the attractors came to be called 'strange'. Now that their properties are understood in terms of the theory of fractals, they are no longer so strange, perhaps.

Automata

There is an amusing childhood game which involves folding a piece of paper a few times and cutting some wedges and arcs along a folded edge. When the paper is unravelled a delightful symmetric pattern is observed. I can remember doing this to create home-made paper doilies for tea parties.

The fact that large-scale order results from a few nicks and cuts is entirely a consequence of a very simple rule concerning the folding of the paper. The home-made doily is an elementary example of how simple rules and procedures can generate complex patterns. Is there a lesson in this for natural complexity?

P. S. Stevens in his book *Patterns in Nature*[1] points out that the growth of biological organisms often appears to be governed by simple rules. In his classic text *On Growth and Form*[2] D'Arcy Thompson demonstrated how many organisms conform to simple geometrical principles. For example, the shapes of skeletons in a wide variety of fish are related by straightforward geometrical transformations. There is thus a hint that complex *global* patterns in nature might be generated by the repeated application of simple *local* procedures.

The systematic study of simple rules and procedures constitutes a branch of mathematics known as games theory. Related to this is a topic known as cellular automata theory. Orginally introduced by the mathematicians John von Neumann and Stanislaw Ulam as a model for self-reproduction in biological systems, cellular automata have been studied by mathematicians, physicists, biologists and computer scientists for a wide range of applications.

A cellular automaton consists of a regular array of sites or cells, for example like a chequerboard, but usually infinite in extent. The array may be one- or two-dimensional. Each cell can be assigned a value of some variable. In the simplest case the variable takes only two values, which can best be envisaged as the cell being either empty or occupied (e.g. by a counter). The state of the system at any time is then specified by listing which cells are occupied and which are empty.

The essence of the cellular automaton is to assign a rule by which the system evolves deterministically in time in a synchronous manner. If a site is given the value o when empty and 1 when occupied, then rules may be expressed in the form of binary arithmetic. To give an example for a one-dimensional array (line of cells), suppose each cell is assigned the new value o (i.e. is designated empty) if its two nearest neighbours are either both empty or both occupied, and is assigned the value 1 (i.e. is filled) if only one neighbour is occupied (see Figure 22). Arithmetically this corresponds to the rule that the new value of each cell is the sum of its nearest neighbours' values modulo 2. The system may be evolved forward in discrete time steps, in a completely automatic and mechanistic way – hence the name automaton. In practice it is helpful and easy to use a computer with graphic display, but the reader may try the procedure as a game using counters or buttons.

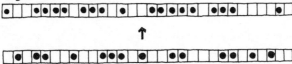

Figure 22. Counters are distributed at random among a line of cells. The system is then evolved forward by one step using the rule described in the text.

The procedure just described is an example of a *local* rule, because the evolution of a given cell depends only on the cells in its immediate vicinity. In all there are 256 possible local rules involving nearest neighbours. The surprise and fascination of cellular automata is that in spite of the fact that the rules are locally defined so there is no intrinsic length scale other than the cell size, nevertheless some automata can spontaneously generate complex large scale patterns displaying long-range order and correlations.

A detailed study of one-dimensional cellular automata has been made by Stephen Wolfram of the Institute for Advanced Study, Princeton. He finds that four distinct patterns of growth emerge. The initial pattern may dwindle and disappear, or simply grow indefinitely at a fixed rate, often generating self-similar forms or fractals, displaying structure on all length scales. Alternatively a pattern may grow and contract in an irregular way, or it may develop to a finite size and stabilize.

Figure 23 shows some examples of the sort of structures that can result from disordered or random initial states. In these cases the system displays the remarkable property of self-organization, a subject to be discussed in depth in the next chapter. Occasionally states of great complexity arise out of featureless beginnings. Wolfram has found states with sequences of periodic structures, chaotic non-periodic behaviour and complicated localized structures that sometimes propagate across the array as coherent objects. Cases of self-reproduction have also been observed. Although different initial states lead to differences of detail in the subsequent patterns, for given rules the same sort of features tend to recur for a wide range of initial states. On the other hand, the behaviour of the automaton differs greatly according to which particular set of rules is applied.

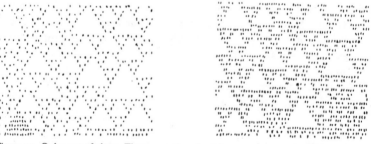

Figure 23. Order out of chaos. These examples of cellular automata start out with randomly distributed counters and spontaneously arrange themselves into ordered patterns with long-range correlations.

More realism can be injected into cellular automata by including the effects of noise, because all natural systems are subject to random fluctuations which would perturb any simple local activity. This can be incorporated into the automaton by replacing its rigidly deterministic rules with a probabilistic procedure. The additional random element can then be analysed statistically. Because of the crucial feedback that is built into the automaton, its evolution differs greatly from the conventional systems studied in statistical mechanics or thermodynamics, which are close to equilibrium (so-called Markovian systems). Disordered or random initial states can evolve definite structures containing long sequences of correlated sites.

Wolfram discovers that:[3]

Starting even from an ensemble in which each possible configuration appears with equal probability, the cellular automaton evolution concentrates the probabilities for particular configurations, thereby reducing entropy. This phenomenon allows for the possibility of self-organization by enhancing the probabilities of organized configurations and suppressing disorganized configurations.

An important property of most cellular automata is that their rules are irreversible, i.e. not symmetric in time. They thus escape from the strictures of the second law of thermodynamics, which is based on reversibility in the underlying microscopic dynamics. For this reason, as quoted above, the entropy of automaton states can decrease, and order can spontaneously appear out of disorder. In this respect cellular automata resemble Prigogine's dissipative structures, to be discussed in Chapter 6, for which the underlying physics is also strongly irreversible, and which develop order out of chaos. Indeed, close analogues of limit cycles and strange attractors are found with some automaton rules.

The hope is that the study of simple automata will uncover new universal principles of order that may be displayed in much more complex natural systems. Wolfram and his colleagues claim:[4]

Analysis of general features of their behaviour may therefore yield general results on the behaviour of many complex systems, and may perhaps ultimately suggest generalizations of the laws of thermodynamics appropriate for systems with irreversible dynamics.

One particular class of cellular automata, called additive, is especially tractable:[5]

Global properties of additive cellular automata exhibit a considerable universality, and independence of detailed aspects of their construction . . . It potentially allows for generic results, valid both in the simple cases which may easily be analyzed, and in the presumably complicated cases which occur in real physical systems.

In the case of two-dimensional arrays (infinite chequerboards) the variety and richness of the generated complexity is much greater. A famous example is provided by the 'game' known as *Life* invented by mathematician John Conway. The in-depth study of this automaton reveals structures that move about coherently, reproduce, undergo life cycles, attack and destroy other structures, and generally cavort about in an intriguing and entertaining way.

It is worth remarking in passing that cellular automata can also be analysed as formal logical systems, and their time evolution viewed in terms of information processing. They can therefore be treated as computers. It has been proved that a certain class of cellular automaton can simulate a so-called Turing machine, or universal computer, and thus be capable of evaluating any computable function, however complex. This could prove of great practical value in designing much sought-after parallel processing computer systems.

Von Neumann argued that Turing's proof of the existence of a universal computing machine could be adapted to prove the possibility of a universal *constructor*, or self-reproducing automaton. We are familiar with man-made machines that make other machines, but the constructors are always more complicated than their products. On the other hand living organisms succeed in making other organisms at least as complicated as themselves. Indeed when evolution is taken into account, the products must sometimes be more complicated than the original.

The question of whether it is possible for a machine equipped with a program to reproduce itself was investigated by von Neumann. Now this amounts to more than simply convincing oneself that a machine can be programmed to make a replica; the replica must itself be capable of self-reproduction. Thus the original machine not only has to make another machine, it also has to make a new set of instructions to enable the other machine to replicate. So the program must contain details of how to make the new hardware plus how to replicate the instructions themselves. There is a danger of an infinite regress here.

It took von Neumann 200 pages of his book *Theory of Self-Reproducing Automata*[6] to prove rigorously that in fact it *is* possible for a universal constructor to exist. He found, however, that self-reproduction can only occur when the machine exceeds a certain threshold of complication. This is a most significant result because it demonstrates that a physical system can take on qualitatively new properties (e.g. self-reproducibility) when it possesses a certain level of complexity.

Video feedback

Video cameras are commonplace pieces of equipment these days. In simple terms a video camera produces an image on a television monitor screen of the scene it is looking at. But what happens when a video camera looks at its own monitor?

The situation has a touch of paradox, reminiscent of Epimenides ('This statement is a lie') and other famous paradoxes of self-reference. As might be expected, when a video camera peers at its own soul, the system goes haywire, as readers who have access to such equipment may easily verify. However, the result is not always a chaotic blur of amorphous shapes. The images show a surprising tendency to develop order and structure spontaneously, turning into pinwheels, spirals, mazes, waves and striations. Sometimes these forms stabilize and persist, sometimes they oscillate rhythmically, the screen flashing through a cycle of colours before returning to its starting form. A self-observing video system is thus a marvellously vivid example of self-organization.

As we shall see in the next chapter, an essential element in all mechanisms of self-organization is feedback. In normal operation, a video camera collects visual information, processes it, and relays it to a remote screen. In the case that the camera inspects its own monitor, the information goes round and round in a loop. The resulting behaviour amounts to more than just a demonstration of self-organization in visual patterns, it is being seriously studied as a test bed for improving understanding of spatial and dynamical complexity in general.

Some researchers believe that video feedback could hold clues about the growth of biological forms (morphogenesis), and also throw light (literally) on the theories of cellular automata, chaotic systems, and chemical self-organization. As James Crutchfield of the Center for Nonlinear Dynamics at Los Alamos explains:[7]

The world about us is replete with complexity arising from its interconnected-ness . . . This interconnectedness lends structure to the chaos of microscopic physical reality that completely transcends descriptions based on our traditional appreciation of dynamical behaviour . . . I believe that video feedback is an intermediary step, a prerequisite for our comprehending the complex dynamics of life.

In practical terms, video feedback is quick and easy to achieve. The camera is set up a few feet in front of the screen in a darkened room. The experimenter has freedom to adjust focus, zoom, brightness, distance and orientation of the camera, all of which affect the nature of the image. To get started the light can be switched on, and a hand waved in front of the camera. Images will start to dance about on the screen, and after a certain amount of trial and error, coherent patterns can be obtained.

Crutchfield has analysed video feedback in detail, and believes the behaviour of the system can be understood in terms that are very familiar in other complex dynamical systems. He suggests that the video system can be described using equations similar to the reaction–diffusion equations used to model chemical self-organization and biological morphogenesis. For example, video feedback is a dissipative dynamical system (see Chapter 6), and the state of the system (represented by the instantaneous image on the screen) can evolve under the influence of *attractors* in direct analogy to, say, the forced pendulum.

It is possible to find direct analogues of non-linear mechanical behaviour. Thus the system may settle down to a stable image, corresponding to a point attractor, or undergo the periodic changes associated with a limit cycle. Alternatively, the state may approach a chaotic (fractal) attractor, leading to unpredictable and erratic behaviour. As in other systems that are progressively driven away from equilibrium, the video system may become unstable at certain critical values of some parameter such as zoom. In this case bifurcations occur, causing the system to jump abruptly and spontaneously into a new pattern of activity, perhaps a state of higher organization and complexity.

All these interesting features make video feedback a fascinating tool for the simulation of complexity and organization in a wide range of physical, chemical and biological systems. It may well be that the video system, in generating pattern and form spontaneously, can elucidate some of the general principles whereby complex structures arise in the natural world.

Is the study of fractals, cellular automata, video feedback and the like merely an amusing diversion, a mimic of natural complexity, or does it address fundamental principles that nature employs in the real world? Superficial similarity can, of course, be beguiling. After all, cartoons can appear very lifelike, but bear no relation to the principles on which real life is based. Computers can respond as though intelligent even when programmed in a very elementary way that is known to have no connection with the way the brain operates.

Proponents of cellular automata point out that many natural systems of great complexity are built out of more or less identical units or components. Biological organisms are made from cells, snowflakes from ice crystals, galaxies from stars, etc. Specific automata patterns have been identified with, for example, pigmentation arrangements on mollusc shells, and spiral galaxies. It is argued that cellular automata provide a tractable and suggestive means of modelling self-organization in a wide range of physical, chemical and biological systems. Perhaps more importantly, the study of cellular automata may lead to the discovery of general principles concerning the nature and generation of complexity when assemblages of simple things act together in a cooperative way. According to Wolfram:[8] 'The ultimate goal is to abstract from a study of cellular automata general features of "self-organizing" behaviour and perhaps to devise universal laws analogous to the laws of thermodynamics'.

The computational and analogical studies reported in this chapter are certainly provocative, and indicate that the appearance of complex organized systems in nature might well comply with certain general mathematical principles. What can be said, then, about the way in which organization and complexity arise spontaneously in nature? This takes us to the subject of real self-organizing systems.

6
Self-organization

Creative matter

Anyone who has stood by a fast flowing stream cannot fail to have been struck by the endlessly shifting pattern of eddies and swirls. The turmoil of the torrent is revealed, on closer inspection, to be a maelstrom of organized activity as new fluid structures appear, metamorphose and propagate, perhaps to fade back into the flow further downstream. It is as though the river can somehow call into fleeting existence a seemingly limitless variety of forms.

What is the source of the river's creative ability?

The conventional view of physical phenomena is that they can ultimately all be reduced to a few fundamental interactions described by deterministic laws. This implies that every physical system follows a unique course of evolution. It is usually assumed that small changes in the initial conditions produce small changes in the subsequent behaviour.

However, now a completely new view of nature is emerging which recognizes that many phenomena fall outside the conventional framework. We have seen how determinism does not necessarily imply predictability: some very simple systems are infinitely sensitive to their initial conditions. Their evolution in time is so erratic and complex that it is essentially unknowable. The concept of a unique course of evolution is then irrelevant. It is as though such systems have a 'will of their own'.

Many physical systems behave in the conventional manner under a range of conditions, but may arrive at a threshold at which predictability suddenly breaks down. There is no longer any unique course, and the system may 'choose' from a range of alternatives. This usually signals an abrupt transition to a new state which may have very different properties. In many cases the system makes a sudden leap to a much more elaborate and complex state. Especially interesting are those cases where spatial patterns or temporal rhythms spontaneously appear. Such states seem to possess a degree of *global cooperation*. Systems which undergo transitions to these states are referred to as *self-organizing*.

Examples of self-organization have been found in astronomy, physics, chemistry and biology. The familiar phenomenon of turbulent flow mentioned already has puzzled scientists and philosophers for millennia. The onset of turbulence depends on the speed of the fluid. At low speed the flow is smooth and featureless, but as the speed is increased a critical threshold occurs at which the fluid breaks up into more complex forms. Further increase in speed can produce additional transitions.

The transition to turbulent flow occurs in distinct stages when a fluid flows past an obstacle such as a cylinder. At low speed the fluid streams smoothly around the cylinder, but as the speed is increased a pair of vortices appears downstream of the obstacle. At higher speeds the vortices become unstable and break away to join the flow. Finally, at yet higher speed the fluid becomes highly irregular. This is full turbulence. As mentioned briefly in Chapter 5, it is believed that fluid turbulence is an example of 'deterministic chaos'. Assuming this is correct, then the fluid has available to it unlimited variety and complexity, and its future behaviour is unknowable. Evidently we have found the source of the river's creativity.

What is organization?

So far I have been rather loose in my use of the words 'order', 'organization', 'complexity', etc. It is now necessary to consider their meanings rather more precisely.

A clear meaning is attached to phrases such as 'a well-ordered society' or 'an ordered list of names'. We have in mind something in which all the component elements act or are arranged in a cooperative, systematic way. In the natural world, order is found in many different forms. The very existence of laws of nature is a type of order which manifests itself in the various regularities of nature: the ticking of a clock, the geometrical precision of the planets, the arrangement of spectral lines.

Order is often apparent in spatial patterns too. Striking examples are the regular latticeworks of crystals and the forms of living organisms. It is clear, however, that the order implied by a crystal is very different from what we have in mind in an organism. A crystal is ordered because of its very simplicity, but an organism is ordered for precisely the opposite reason – by virtue of its complexity. In both cases the concept of order is a global one; the orderliness refers to the system as a whole. Crystalline order concerns the way that the atomic arrangement repeats itself in a regular pattern throughout the material. Biological order is recognized because the diverse component parts of an organism cooperate to perform a coherent unitary function.

There seems to be an inescapable subjective element involved in the attribution of order. In cryptography, a coded message that is perceived as a disordered and meaningless jumble of symbols by one person might be interpreted as a very carefully constructed document by another. Similarly, a casual glance at an ant-heap might give the impression of a chaotic frenzy, but closer scrutiny would reveal a highly organized pattern of activity.

One way to introduce more objectivity is to choose a mathematical definition to quantify similarity of form. This can be achieved using the concept of *correlations*. For example, different regions of a crystal lattice are highly correlated, people's faces are moderately correlated, clouds are usually only very weakly correlated in shape. It is possible to make such comparisons mathematically precise, and even to turn the search for correlations over to computers. Sometimes automatic searches can uncover correlations where none were perceived before, as in astronomy, where photographs of apparently haphazard distributions of galaxies can be shown to contain evidence for clustering.

Using mathematics one also obtains a definition of randomness, which is often regarded as the opposite of order. For instance, a random sequence of digits is one in which no systematic patterns of any kind exist. Note that this does *not* mean that no patterns exist. If a random sequence is searched long enough, the series 1,2,3,4,5 would certainly appear. The point is that its appearance could not be foreseen from examination of the digits that came before. Randomly varying physical quantities are therefore described as chaotic, erratic or disordered.

It is here that we encounter a subtlety. Most computers possess a 'random number generator' which produces, without throwing any dice, numbers that seem to have all the properties of being random. In fact, these numbers are produced from a strictly 'handle-turning' deterministic procedure (e.g. in principle the decimal expansion of π could be calculated, but it would prove to be very time consuming). If one knows the procedure then the sequence becomes exactly predictable, and thus in some sense ordered. Indeed, computers usually give the same sequence of numbers each time they are asked to do so afresh. We might call this 'simulated randomness'. (The technical term is pseudorandomness.) This raises the interesting question of how we could ever tell a pseudorandom sequence from a truly random one if we only have the numbers to work with. How can we know whether the toss of a coin or the throw of a die is truly random? There is no consensus on this. Many scientists believe that the only truly random processes in nature are those of quantum mechanical origin.

It is actually surprisingly hard to capture the concept of randomness mathematically. Intuitively one feels that a random number is in some sense a number without any remarkable or special properties. The problem is, if one is able to define such a number, then the very fact that one has identified it already makes it somehow special. On strategy to circumvent this difficulty is to describe numbers algorithmically, that is, in terms of the output of some computer program. We have already met this idea at the end of Chapter 3 in connection with the subject of the jumping particle. Special (i.e. nonrandom) numbers are then those numbers that can be generated by a program containing fewer bits of information than the number itself. A random number is then a number that cannot be thus generated. It turns out, using this definition of randomness, that almost all numbers are random, but that most of them cannot be *proved* to be random!

Physicists and chemists quantify order in a way that relates to entropy, as already discussed in Chapter 2. In this case true disorder corresponds to thermodynamic equilibrium. It is important to realize, however, that this definition refers to the molecular level. A flask uniformly filled with liquid at a constant temperature is essentially featureless to the naked eye, even though it may have attained maximum entropy. It doesn't seem to be *doing* anything disorderly! But things would look very different if we could view the molecules rushing about chaotically. By contrast, the boiling contents of a kettle may appear to be disorderly, which in a *macroscopic* sense they are, but from the thermodynamic point of view this system is not in equilibrium, so it is not maximally disordered. There may be order on one scale and disorder on another.

Very often order is used interchangeably with organization, but this can be misleading. It is natural to refer to a living organism as organized, but this would not apply to a crystal, though both are ordered. Organization is a quality that is most distinctive when it refers to a process rather than a structure. An amoeba is organized because its various components work together as part of a common strategy, each component playing a specialized and interlinking role with the others. A fossil may retain something of the form of an organism, and is undeniably ordered, but it does not share the organization of its originator, because it is 'frozen'. The distinction between order and organization can be very important. When bacteria are grown in a culture the total system becomes more organized. On the other hand the second law of thermodynamics requires the total entropy to rise, so in this sense the system as a whole becomes less ordered. It might be said that order refers to the *quantity* of information (i.e. negative entropy) in a system, whereas organization refers to the *quality* of information. When living organisms develop they improve the quality of their environment, but generate entropy in the process.

One can, perhaps, refer to the coming-into-being of order as an example of organization. Thus scientists often talk about the solar nebula organizing itself into a planetary system, or clouds organizing themselves into patterns. This is a somewhat metaphorical usage though, because organization is often taken to imply an element of purpose or design. Still, one gets away with talking about water 'trying to find its own level' and computers 'working out the answer'.

Is it possible to quantify organization or complexity? One obvious difficulty is that neither organization nor complexity is likely to be an additive quantity. By this I mean that we would not regard two bacteria as being twice as complex (or twice as organized) as one, for given one bacterium it is a relatively simple matter to produce two – one merely needs supply some nutrient and wait. Nor is it at all obvious how one would compare the relative degree of complexity of a bacterium with that of, say, a multinational company.

John von Neumann tried to quantify how complicated a system is in terms of the equations describing it. Intuitively, one feels that complexity must take account of the number of components and the richness of their interconnections in some way. Alternatively, and perhaps equivalently, complexity must somehow relate to the information content of the system, or the length of the algorithm which specifies how to construct it. There is a substantial literature of attempts by mathematicians and computer scientists to develop a theory of complexity along these lines.

Charles Bennett, a computer scientist working for IBM, has proposed a definition of organization or complexity based on a concept that he calls 'logical depth'. To take an analogy, suppose A wishes to send a message to B. The purpose might be to communicate the orbit of a satellite, say. Now A could list the successive positions of the satellite at subsequent times. Alternatively, A could simply specify the position and velocity of the satellite at some moment and leave it to B to work out the orbit himself. The latter form of message has all the information content of the former, but it is far less *useful*. In other words, there is more to a message than merely its information content; there is also the *value* or *quality* of the information that has to be taken into account. In this case the logical depth is identified with, roughly speaking, the length of the computation needed to decode the message and reconstruct the orbit. In the case of a physical system, a measure of the logical depth might be the length of computer time needed to simulate the behaviour of the system to some degree of resolution. Bennett has demonstrated how this idea can be formulated in a way that is machine-independent.

A quite different approach to complexity has been followed by the theoretical biologist Robert Rosen of Dalhousie University, Nova Scotia, who stresses that a key characteristic of complex systems is that we can interact with them in a large variety of ways. He thus explicitly recognizes the subjective quality that is inevitably involved. It is not so much what a system is that makes it complex, but what it does. We therefore come up against a teleological element, in which complexity has a purpose. This is obvious in biology, of course. The organized complexity of the eye is for the purpose of enabling the organism to see. It is less obvious how purpose applies to inorganic systems.

One of the important discoveries to emerge from the study of complex systems is that self-organization is closely associated with chaos of the sort discussed in Chapter 4. In one sense chaos is the opposite of organization, but in another they are similar concepts. Both require a large amount of information to specify their states and, as we shall see, both involve an element of unpredictability. The physicist David Bohm has emphasized that complicated or erratic behaviour should not be regarded as disorderly. Indeed, such behaviour requires a great deal of information to specify it, whereas disorder in the thermodynamic sense is associated with the absence of information. Bohm even insists that randomness represents a type of order.

It will be clear from this discussion that organization and complexity, in spite of their powerful intuitive meanings, lack generally agreed definitions in the rigorous mathematical sense. It is to be hoped that as complex systems come to be understood in greater detail this defect will be remedied.

A new kind of order

In 1984, workers at the US National Bureau of Standards discovered a strange material that seemed to possess a new sort of order. Hithero dismissed as an impossibility by scientists, the substance is a solid that displays the same type of order associated with a crystal, except for one important difference. It appears to have symmetries that violate a fundamental theorem of crystallography: its atoms are arranged in a pattern that is physically impossible for any crystalline substance. It has therefore been dubbed a *quasicrystal*.

A normal crystal is a latticework of atoms arrayed in a highly regular repeating pattern. The various crystalline forms can be classified using the mathematical theory of symmetry. For example, if the atoms occupy sites corresponding to the corners of a cube, the lattice has four-fold rotational symmetry because it would look the same if rotated by one-quarter of a revolution. The cube can be considered as the unit building block of the lattice, and one can envisage a space-filling collection of cubes fitting together snugly to form a macroscopic lattice.

The rules of geometry and the three-dimensionality of space place strong restrictions on the nature of crystal symmetries. A simple case that is ruled out is *five-fold* rotational symmetry. No crystalline substance can be five-fold symmetric.

The reason is simple. Everybody has seen a wall completely tiled with squares (four-fold symmetry). It can also be done with hexagons (six-fold symmetry) as the bees have discovered. But nobody has seen a wall completely tiled with pentagons, because it can't be done. Pentagons don't fit together without leaving gaps (see Figure 24).

In three dimensions the role of the pentagon is played by a five-fold symmetric solid with the fearsome name of icosahedron, a figure with 20 triangular faces arranged such that five faces meet at each vertex. Whilst you could pack a crate with cubes leaving no spaces, you would try in vain to do the same with icosahedra. They simply cannot be snugly fitted

Figure 24. Tesselating pentagons is impossible: they won't fit snugly.

together in a space-filling way. This means that while an individual group of atoms might be arranged in the shape of an icosahedron, it is impossible for a periodic latticework of such units to be constructed. For this reason it came as a shock when electron microscope studies at the National Bureau of Standards revealed large-scale five-fold symmetry in an alloy of aluminium and manganese.

At this stage a number of people began to take note of a curious discovery made several years earlier by Oxford mathematician Roger Penrose, who is better known for his work on black holes and space-time singularities. Penrose showed how it is possible to tile the plane with five-fold symmetry using two shapes, a fat rhombus and a thin rhombus. The resulting pattern is shown in Figure 25. The pentagonal symmetry is apparent in the many decagons (ten-sided figures) that can be found. There is clearly a degree of long-range order, because the sides of the decagons are parallel to one another.

To understand the essential difference between the Penrose tiling and a crystalline pattern, one must distinguish between two sorts of long-range order, translational and orientational. Both are possessed by conventional crystals with periodic lattices. Translational order refers to the fact that the lattice would look the same if it was shifted sideways by one building unit (e.g. one cube), or any exact multiple thereof. Orientational order is the property that the unit building blocks of atoms form geometrical figures whose edges and faces are oriented parallel to each other throughout the crystal.

Penrose's tiling pattern, which serves as a model for quasicrystals, possesses orientational but not translational order. It evades the theorem that precludes pentagonal symmetry because, unlike a crystal lattice, it is not periodic: however far the tiling is extended, no local pattern will ever recur cyclically.

How should such patterns be described? They undeniably possess a simple form of holistic order, but there is also a high degree of complexity because the pattern is everywhere slightly different. This raises the baffling question of how quasicrystals grow in the first place. In a conventional crystal the order present in a unit building block propagates across the whole lattice by simple repetition. The physical forces acting on corresponding atoms in the different blocks are the same everywhere. In a quasi-crystal each five-fold block sits in a slightly different environment, with a different pattern of forces. How do the atoms of the different elements present conspire to aggregate in the right proportions and in the correct locations to maintain orientational order over such long distances,

Figure 25. Penrose's tiling pattern. Using only two shapes the entire plane can be covered without gaps to produce a remarkable pattern that has five-fold symmetry and long-range order, but **no** periodicity.

when each atom is subject to different forces? There seems to be some sort of non-local organizating influence that is as yet a complete mystery.

Examples of self-organization

The simplest type of self-organization in physics is a phase transition. The most familiar phase transitions are the changes from a liquid to a solid or a gas. When water vapour condenses to form droplets, or liquid water freezes to ice, an initially featureless state abruptly and spontaneously acquires structure and complexity.

Phase transitions can take many other forms too. For example, a ferromagnet at high temperature shows no permanent magnetization, but as the temperature is lowered a critical threshold is reached at which magnetization spontaneously appears. The ferromagnet consists of lots of microscopic magnets that are partially free to swivel. When the material is hot these magnets are jiggled about chaotically and independently, so that on a macroscopic scale their magnetizations average each other out. As the material is cooled, the mutual interactions between the micromagnets try to align them. At the critical temperature the disruptive effect of the thermal agitation is suddenly overcome, and all the micromagnets cooperate by lining up into an ordered array (see Figure 26). Their magnetizations now reinforce to produce a coherent large scale field.

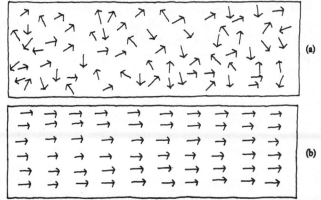

Figure 26. (a) At high temperatures thermal agitation keeps the micro-magnets oriented randomly: their fields average to zero. (b) Below a critical temperature a phase transition occurs, and all the micro-magnets spontaneously organize themselves into a coherent pattern. The long-range order resulting from this cooperative behaviour ensures that the component magnetic fields add together to produce a macroscopic field.

Another example concerns electrical conductivity. When certain substances are cooled to near absolute zero, they suddenly lose all resistance to electricity and become *superconducting*. In this low temperature phase, the billions of electrons that constitute the current behave as one, moving in a highly correlated and organized quantum wave pattern. This is in contrast to the situation in an ordinary conductor, where the electrons move largely independently and follow complicated and erratic paths. Similar large-scale organization occurs in *superfluids*, such as liquid helium, where the fluid can flow without friction.

The foregoing examples of self-organization occur when the temperature is gradually lowered under conditions of thermodynamic equilibrium. More dramatic possibilities arise when a system is driven far away from equilibrium. One such case is the laser. Near to thermodynamic equilibrium a hot solid or gas behaves like an ordinary lamp, with each atom emitting light randomly and independently. The resulting beam is an incoherent jumble of wave trains each a few metres long. It is possible to drive the system away from equilibrium by 'pumping', which is a means of giving energy to the atoms to put an excessive number of them into excited states. When this is done a critical threshold is reached at which the atoms suddenly organize themselves on a global scale and execute cooperative behaviour to a very high level of precision. Billions of atoms emit wavelets that are exactly in phase, producing a coherent wave train of light that stretches for thousands of miles.

Another example of spontaneous self-organization in a system driven far from equilibrium is the so-called Bénard instability, which occurs when a horizontal layer of fluid is heated from below. As explained briefly in Chapter 4, this is the situation in meteorology where sunlight heats the ground, which then heats the air above it. It also occurs in every kitchen when a pan of water is placed on a stove. The warm liquid near the base is less dense and tries to rise. So long as the temperature difference between the top and bottom of the liquid is small (near to equilibrium) the upthrust is resisted by viscosity. As the base temperature is raised, however, a threshold is crossed and the liquid becomes unstable; it suddenly starts to convect. Under carefully regulated conditions, the convecting liquid adopts a highly orderly and stable pattern of flow, organizing itself into distinctive rolls, or into cells with a hexagonal structure. Thus an initially homogeneous state gives way to a spatial pattern with distinctive long-range order. Further heating may induce additional transitions, such as the onset of chaos.

An important feature in all these examples is that a *symmetry*, present initially, is broken by the transition to a more complex phase. Take the case of water freezing to ice. A homogeneous volume of water possesses rotational symmetry. When ice crystals form, the symmetry is lost because the crystal planes define a preferred orientation in space.

Symmetry breaking also occurs in the transition to ferromagnetism. The high-temperature phase also has rotational symmetry, because the micromagnets average their magnetic fields over all orientations. When the temperature falls, the micromagnets align, again defining a preferred spatial direction and breaking the rotational symmetry.

These are examples of geometrical symmetry breaking. Modern particle physics makes use of more generalized symmetry concepts, such as abstract *gauge* symmetries, which can also become spontaneously broken. Because symmetries are generally broken as the temperature is lowered, the history of the universe, in cooling from the very hot initial phase, is a succession of symmetry breaks. Symmetry breaking thus provides an alternative to complexity as a measure of the universe's progressive creative activity.

Dissipative structures: a theory of form

True scientific revolutions amount to more than new discoveries; they alter the concepts on which science is based. Historians will distinguish three levels of enquiry in the study of matter. The first is Newtonian mechanics – the triumph of necessity. The second is equilibrium thermodynamics – the triumph of chance. Now there is a third level, emerging from the study of far-from-equilibrium systems.

Self-organization occurs, as we have seen, both in equilibrium and non-equilibrium systems. In both cases the new phase has a more complex spatial form. There is, however, a fundamental difference between the type of structure present in a ferromagnet and that in a convection cell. The former is a static configuration of matter, frozen in a particular pattern. The latter is a dynamical entity, generated by a continual throughput of matter and energy from its environment: the name *process structure* has been suggested.

It is now recognized that, quite generally, systems driven far from equilibrium tend to undergo abrupt spontaneous changes of behaviour. They may start to behave erratically, or to organize themselves into new and unexpected forms. Although the onset of these abrupt changes can sometimes be understood on theoretical grounds, the detailed form of the new phase is essentially unpredictable. Observing convection cells, the physicist can explain, using traditional concepts, why the original homogeneous fluid became unstable. But he could not have predicted the detailed arrangement of the convection cells in advance. The experimenter has no control over, for example, whether a given blob of fluid will end up in a clockwise or anticlockwise rotating cell.

A crucial property of far-from-equilibrium systems that give rise to process structures is that they are *open* to their environment. Traditional techniques of physics and chemistry are aimed at closed systems near to equilibrium, so an entirely new approach is needed. One of the leading figures in developing this new approach is the chemist Ilya Prigogine. He prefers the term *dissipative structure* to describe forms such as convection cells.

To understand why, think about the motion of a pendulum. In the idealized case of an isolated frictionless pendulum (closed system), the bob will swing for ever, endlessly repeating the same pattern of motion. If the pendulum is jogged, the motion adopts a new pattern which is permanently retained. One could say that the pendulum remembers the disturbance for all time.

The situation is very different if friction is introduced. The moving pendulum now dissipates energy in the form of heat. Whatever its initial motion, it will inexorably come to rest. (This was described in Chapter 4 as the representative point converging on a limit point in the phase diagram.) Thus, it loses all memory of its past history.

If the damped pendulum is now driven by a periodic external force it will adopt a new pattern of motion dictated by that force. (This is limit cycle behaviour.) We might say that the ordered motion of the pendulum is imposed by a new *organizing agency*, namely the external driving force. Under these circumstances the orderly activity of the system is stable (assuming there are no non-linear effects leading to chaos). If the pendulum is perturbed in some way, it soon recovers and settles back to its former pattern of motion, because the perturbation is damped away by the dissipation. Again, the memory of the disturbance is lost.

The driven damped pendulum is a simple example of a dissipative structure, but the same principles apply quite generally. In all cases the system is driven from equilibrium by an external forcing agency, and it adopts a stable form by dissipating away any perturbations to its structure. Because energy is continually dissipated, a dissipative structure will only survive so long as it is supplied with energy (and perhaps matter too) by the environment.

This is the key to the remarkable self-organizing abilities of far-from-equilibrium systems. Organized activity in a closed system inevitably decays in accordance with the second law of thermodynamics. But a dissipative structure evades the degenerative effects of the second law by exporting entropy into its environment. In this way, although the total entropy of the universe continually rises, the dissipative structure maintains its coherence and order, and may even increase it.

The study of dissipative structures thus provides a vital clue to understanding the generative capabilities of nature. It has long seemed paradoxical that a universe apparently dying under the influence of the second law nevertheless continually increases its level of complexity and organization. We now see how it is possible for the universe to increase both organization and entropy at the same time. The optimistic and pessimistic arrows of time can co-exist: the universe can display creative unidirectional progress even in the face of the second law.

The chemical clock

Prigogine and his colleagues have studied many physical, chemical and biochemical dissipative processes which display self-organization. A very striking chemical example is the so-called Belousov-Zhabatinski reaction. A mixture of cerium sulphate, malonic acid and potassium bromate is dissolved in sulphuric acid. The result is dramatic.

In one experiment a continual throughput of reagents is maintained using pumps (note the essential openness again), and the system is kept well stirred. To keep track of the chemical condition of the mixture, dyes can be used which show red when there is an excess of Ce^{3+} ions and blue when there is an excess of Ce^{4+} ions. For low rates of pumping (i.e. close to equilibrium) the mixture remains in a featureless steady state. When the throughput is stepped up, however, forcing the system well away from equilibrium, something amazing happens. The liquid suddenly starts to turn blue throughout. This lasts for a minute or two. Then it turns red, then blue, then red, and so on, pulsating with perfect regularity. Prigogine refers to this remarkable rhythmic behaviour as a *chemical clock*.

It is important to appreciate the fundamental distinction between this chemical clock and the rhythmic swinging of a simple pendulum. The pendulum is a system with a single degree of freedom, and it executes oscillations in the absence of dissipation. If dissipation is present then, as discussed above, the regular periodic motion has to be imposed by an external driving force. By contrast, the chemical clock has a vast number (10^{23}) of degrees of freedom, and is a dissipative system. Nevertheless the pulsations are not imposed by the external forcing agency (the forced input of reagents) but are produced by an *internal* rhythm of some sort, that depends on the dynamical activity of the chemical reaction.

The explanation of the chemical clock can be traced to certain chemical changes that take place in the mixture in a cyclic fashion, with a natural frequency determined by the concentrations of the various chemicals involved. An essential element in this cyclic behaviour is the phenomenon of 'autocatalysis'. A catalyst is a substance that accelerates a chemical reaction. Autocatalysis occurs when the presence of a substance promotes the further production of that same substance. Engineers call this sort of thing feedback. In mathematical terms autocatalysis introduces *non-linearity* into the system. The result, as ever, is a form of symmetry breaking. In this case the initial state is symmetric under time translations (it looks the same from one moment to the next), but this symmetry is spontaneously broken by the oscillations.

The really surprising feature of the Belousov–Zhabatinski reaction is the degree of coherence of the chemical pulsations. After all, chemical reactions take place at the molecular level. The forces between individual molecules have a range of only about a ten-millionth of a centimetre. Yet the chemical clock displays orderly behaviour over a scale of *centimetres*. Countless trillions of atoms cooperate in perfectly synchronous behaviour, as though subordinated to a sort of global plan.

Alvin Tofler, in a foreword to one of Prigogine's books, describes this bizarre phenomenon as follows:[1]

> Imagine a million white ping-pong balls mixed at random with a million black ones, bouncing around chaotically in a tank with a glass window in it. Most of the time the mass seen through the window would appear to be gray, but now and then, at irregular moments, the sample seen through the glass might seem black or white, depending on the distribution of the balls at that moment in the vicinity of the window.
>
> Now imagine that suddenly the window goes all white, then all black, then all white again, and on and on, changing its colour completely at fixed intervals – like a clock ticking.
>
> Why do all the white balls and all the black ones suddenly organize themselves to change colour in time with one another? By all the traditional rules this should not happen at all.

Traditional chemistry, of course, deals with systems close to equilibrium. Yet equilibrium conditions are highly idealized and rarely found in nature. Nearly every naturally occurring chemical system is far from equilibrium, and this regime remains largely unexplored. But clearly, as in the case of simple physical systems, far-from-equilibrium chemical systems are likely to show surprising and unpredictable behaviour.

The Belousov–Zhabatinski reaction resembles in many ways the motion of a simple dynamical system, where the chemical concentrations play the role of dynamical variables. One can discuss the reaction pictorially using phase diagrams, trajectories, limit cycles, etc. as before. Using this language, the reaction discussed here can be viewed as a limit cycle, similar to the driven pendulum. As in that case, if the chemical forcing is increased, the simple rhythmic behaviour of the chemical mixture gives way to more and more complex oscillatory patterns, culminating in chaos – large scale chemical chaos, not the molecular chaos associated with thermodynamic equilibrium.

As well as long-range temporal order, the Belousov–Zhabatinski reaction can display long-range spatial order. This comes about if the reagents are arranged in a thin layer and left unstirred. Various geometrical wave forms then spontaneously appear and grow in the mixture. These can take the shape of circular waves that emanate from certain centres and expand at fixed speed, or spirals that twist outwards either clockwise or anticlockwise. These shapes provide a classic example of the spontaneous appearance of complex forms from an initially featureless state, i.e. spatial symmetry breaking. They are the spatial counterpart of the temporal symmetry breaking displayed in the chemical clock.

Matter with a 'will of its own'

It is hard to overemphasise the importance of the distinction between matter and energy in, or close to, equilibrium – the traditional subject for scientific study – and far-from-equilibrium dissipative systems. Prigogine has referred to the latter as *active matter*, because of its potential to spontaneously and unpredictably develop new structures. It seems to have 'a will of its own'. Disequilibrium, claims Prigogine, 'is the source of order' in the universe; it brings 'order out of chaos'.

It is as though, as the universe gradually unfolds from its featureless origin, matter and energy are continually being presented with alternative pathways of development: the passive pathway that leads to simple, static, inert substance, well described by the Newtonian or thermodynamic paradigms, and the active pathway that transcends these paradigms and leads to unpredictable, evolving complexity and variety. 'In the modern world view,' writes Charles Bennett,[2] 'dissipation has taken over one of the functions formerly performed by God: it makes matter transcend the clod-like nature it would manifest at equilibrium, and behave instead in dramatic and unforeseen ways, molding itself for example into thunderstorms, people and umbrellas'.

The appearance of diverging pathways of evolution is in fact a very general feature of dynamical systems. Mathematically, the situation can be described using so-called partial differential equations. These equations can only be solved by specifying boundary conditions for the system. In the case of open systems, the external world exercises a continual influence through the boundaries in the form of unpredictable *fluctuations*. Examination of the solutions of the equations reveals the general feature that, for systems close to equilibrium, fluctuations are suppressed. As the system is forced farther and farther from equilibrium, however, the system reaches a critical point, technically known as a bifurcation point. Here the original solution of the equations becomes unstable, signalling that the system is about to undergo an abrupt change.

The situation is depicted schematically in Figure 27. The single line represents the original equilibrium solution, which then branches, or bifurcates at some critical value of a physical parameter (e.g. the temperature difference between the top and bottom of a fluid layer). At this point the system has to choose between the two pathways. Depending

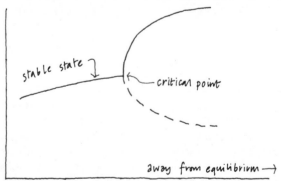

Figure 27. Nature has 'free will'. The graph illustrates how, when a physical system is driven progressively further from equilibrium, a unique state may suddenly become unstable, and face two alternative pathways of evolution. No prediction can be made about which 'branch' will be chosen. Mathematically, the single line is a solution to the evolution equations which bifurcates at a singularity in the equations that occurs when a forcing parameter reaches some critical value.

on the context this may be the moment when the system leaps into a new state of enhanced organization, developing a novel and more elaborate structure. Or it may instead become unstable and descend into chaos. At the bifurcation point the inescapable fluctuations, which in ordinary equilibrium thermodynamics are automatically suppressed, instead become amplified to macroscopic proportions, and drive the system into its new phase which then becomes stabilized.

Because the system is open, the form of these endless microscopic fluctuations is completely unknowable. There is thus an inherent uncertainty in the outcome of the transition. For this reason, the detailed form of the new organized structures is intrinsically unpredictable. Prigogine calls this phenomenon *order through fluctuations*, and proposes that it is a fundamental organizing principle in nature:[3] 'It seems that environmental fluctuations can both *affect bifurcation* and – more spectacularly – *generate new nonequilibrium transitions* not predicted by the phenomenological laws of evolution.'

A very simple example of a bifurcation is provided by the case of a ball at rest in equilibrium at the bottom of a one-dimensional valley (see Figure 28). Suppose this system is disturbed by a symmetric upthrust of the valley, which carries the ball vertically. Initially friction prevents the ball from rolling, but as the distance from equilibrium increases, the instability of the ball gets more and more precarious until, at some critical point, the ball rolls off the hump and into the valley. At this moment the solution of the mechanical equations bifurcates into two branches, representing two new stable minimum energy states.

Once more we encounter symmetry breaking. The original configuration, which was symmetric, gives way to a lopsided arrangement: symmetry is traded for stability. The ball may choose either the right-hand or left-hand valleys. Which it will choose depends, of course, on the microscopic fluctuations that may jog it a minute distance one way or the other. This microscopic twitch is then amplified and the ball is sent rolling into one of the valleys at an accelerating rate. By their very nature the microscopic fluctuations are unpredictable, yet they are ultimately responsible for driving the system into a completely different macroscopic state.

The simple example given above is a case of *static* equilibrium, but similar ideas carry over to dynamic processes such as limit cycles and dissipative structures. An example of bifurcation in a dynamical process is shown in Figure 29. Here a rigid rod with a mass attached to the end forms a pendulum and is free to rotate in a plane. For low values of energy,

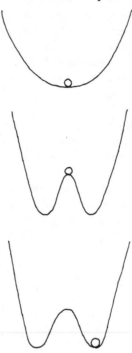

Figure 28. When a hump rises beneath the ball the symmetric state becomes unstable. The ball rolls into one or other valley, spontaneously breaking the symmetry. This represents a bifurcation at which the ball can 'choose' between two contending configurations.

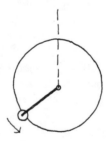

Figure 29. For low values of the energy the pendulum swings back and forth. As the energy rises (i.e. the system is driven further from equilibrium) the swings become bigger until, at a critical value, the arm crosses the vertical line and oscillation becomes rotation. The system has suddenly flipped to a completely new pattern of behaviour.

the pendulum swings backwards and forwards in the traditional way. As the energy is increased, the swings get higher and higher, until a critical value is reached at which the motion just carries the pendulum bob to the top of the circle. At this value the nature of the motion changes abruptly and drastically. Instead of oscillating, the rod travels across the vertical and falls down the other side. Oscillation has given way to rotation. The pendulum has suddenly altered its activity to a completely new pattern.

How can these abrupt changes in behaviour be understood mathematically? Many examples of chemical self-organization can be successfully modelled using something known as the reaction–diffusion equation. This equation expresses the rate of change of the concentration of a particular chemical as a sum of two factors. The first represents the increase or decrease in the amount of the chemical as a result of chemical reactions among the other substances participating in the scheme. The second arises because in a real system chemicals diffuse into their surroundings, and this will alter the concentrations in different regions. It turns out that this simple equation can describe a quite extraordinary range of behaviour, including the key features of instabilities and bifurcations. It can lead to changes in time such as the chemical clock, and to spatial forms such as the spiral waves of the Belousov–Zhabatinski reaction.

One of the earliest analyses along these lines was carried out by the mathematician Alan Turing in 1952. Turing is perhaps best known for his epochal work on the foundations of mathematics, especially in relation to the concept of a universal computing machine already mentioned briefly in Chapter 5. Turing's wartime work at Bletchley Park on cracking the German 'Enigma' code, which resulted in the saving of many Allied lives, led to the construction of the first real computer and turned Turing into a legendary figure. His suicide in 1953 deprived science of one of its finer intellects.

Turing combined his fascination for the foundations of mathematics with a lively interest in biology, in particular in the appearance of certain forms in plants and animals that were suggestive of geometrical patterns. By what mechanism, wondered Turing, do these forms arrange themselves?

As a simple example of the sort of processes that could be responsible, Turing considered what would happen to, say, two chemical substances that could enhance or inhibit each other's production rates, and could also diffuse into their surroundings. Turing was able to demonstrate mathematically that, for certain values of the diffusion and reaction rates, a wave of chemical concentration could develop. If it is supposed that the concentration of the chemicals somehow triggers the onset of growth, then it is possible to envisage the establishment of a sort of chemical framework providing the positional information that tells the organism where and how to grow. In this way, chemical patterns of the Belousov–Zhabatinski type might conceivably play a role in biological morphogenesis.

One of the fascinating aspects of the work of Prigogine and others on dissipative structures is that a common language is developed for the description of both living and non-living – indeed quite ordinary – systems. Concepts such as coherence, synchronization, macroscopic order, complexity, spontaneous organization, adaptation, growth of structure and so on are traditionally reserved for biological systems, which undeniably have 'a will of their own'. Yet we have been applying these terms to lasers, fluids, chemical mixtures and mechanical systems. The third level of enquiry into matter is transforming our understanding of nature's most conspicuous manifestation of self-organization – the phenomenon of life.

7
Life: its Nature

The successes of molecular biology are so beguiling that we forget the organism and its physiology. Schrödinger's disciples, who founded the church of molecular biology, have turned his wisdom into the dogma that life is self-replicating and corrects its errors by natural selection. There is much more to life than this naive truth, just as there is more to the Universe than atoms alone – grandmothers live and enjoy the shade of Lombardy poplar trees not knowing that they and the trees are deemed by this dogma to be dead.

James Lovelock[1]

What is life?

When the quantum physicist Erwin Schrödinger published his little book *What is Life?* in 1944, the title reflected the fact that both the origin and the nature of life seemed deeply mysterious to him. The drift of Schrödinger's thinking proved immensely influential in the rise of the science of molecular biology that soon followed. Nevertheless, in spite of decades of extraordinary progress in unravelling the molecular basis of life, Schrödinger's question remains unanswered. Biological organisms still seem utterly perplexing to scientists.

The problems of understanding life are exemplified by the problems of even defining it. We usually recognize a biological organism as such when we encounter it, yet it is notoriously hard to pin down exactly what it is that makes us so certain something is living. No simple definition will suffice. Any particular property of living systems can also be found in non-living systems: crystals can reproduce, clouds can grow, etc. Clearly, life is characterized by a constellation of unusual properties.

Among the more important features of living things are the following:

Complexity

The degree of complexity in living organisms far exceeds that of any other familiar physical system. The complexity is hierarchical, ranging from the elaborate structure and activity of macromolecules such as proteins and nucleic acid to the exquisitely orchestrated complexity of animal behaviour. At every level, and bridging between levels, is a bewildering network of feedback mechanisms and controls.

Organization

Biological complexity is not merely complication. The complexity is organized and harmonized so that the organism functions as an integrated *whole*.

Uniqueness

Every living organism is unique, both in form and development. Unlike in physics, where one usually studies *classes* of identical objects (e.g. electrons), organisms are individuals. Moreover, collections of organisms are unique, species are unique, the evolutionary history of life on Earth is unique and the entire biosphere is unique. On the other hand, we can recognize a cat as a cat, a cell as a cell, and so on. There are definite regularities and distinguishing features that permit organisms to be classified. Living things seem to be both special and general in a rather precise way.

Emergence

Biological organisms most exemplify the dictum that 'the whole is greater than the sum of its parts'. At each new level of complexity in biology new and unexpected qualities appear, qualities which apparently cannot be reduced to the properties of the component parts.

Holism

A living organism consists of a large range of components, perhaps differing greatly in structure and function (e.g. eyes, hair, liver). Yet the components are arranged and behave in a coherent and cooperative fashion as though to a common agreed plan. This endows the organism with a discrete identity, and makes a worm a worm, a dog a dog, and so forth.

Unpredictability

Although many biological processes are essentially automatic and mechanical, we cannot predict the future state of a biological system in detail. Organisms – especially higher organisms – seem to possess that intriguing 'will of their own'. Moreover, the biosphere as a whole is unpredictable, as evolution throws up novel and unexpected organisms. Cows, ants and geraniums were in no sense inevitable products of evolution.

Openness, interconnectedness and disequilibrium

No living thing exists in isolation. All organisms are strongly coupled to their inanimate environment and require a continual throughput of matter and energy as well as the ability to export entropy. From the physical and chemical point of view, therefore, each organism is strongly out of equilibrium with its environment. In addition, life on Earth is an intricate network of mutually interdependent organisms held in a state of dynamic balance. The concept of life is fully meaningful only in the context of the entire biosphere.

Evolution

Life as we know it would not exist at all unless it had been able to evolve from simple origins to its present complexity. Once again, there is a distinct progression or *arrow of time* involved. The ability of life to evolve and adapt to a changing environment, to develop ever more elaborate structures and functions, depends on its ability to transmit genetic information to offspring (reproduction) and the susceptibility of this information to discrete changes (mutation).

Teleology (or teleonomy)

As noted by Aristotle, organisms develop and behave in an ordered and purposive way, as though guided towards a final goal in accordance with a preordained plan or blueprint. The nineteenth-century physiologist Claude Bernard expressed it thus:[2]

There is, so to speak, a pre-established design of each being and of each organ of such a kind that each phenomenon by itself depends upon the general forces of nature, but when taken in connection with the others it seems directed by some invisible guide on the road it follows and led to the place it occupies.

The modern Nobel biologist and Director of the Pasteur Institute, Jacques Monod, although a strong reductionist, nevertheless makes a similar observation:[3]

> One of the fundamental characteristics common to all living beings without exception [is] that of being *objects endowed with a purpose or project*, which at the same time they show in their structure and execute through their perform- ances . . . Rather than reject this idea (as certain biologists have tried to do) it must be recognized as essential to the very definition of living beings. We shall maintain that the latter are distinct from all other structures or systems present in the universe by this characteristic property, which we shall call *teleonomy*.

Living organisms are the supreme example of active matter. They represent the most developed form of organized matter and energy that we know. They exemplify all the characteristics – growth, adaptation, increasing complexity, unfolding of form, variety, unpredictability – that have been explored in the foregoing chapters. These properties are so prominently represented in living organisms, it is small wonder that the simple question 'What is life?' has led to enormous controversy, and prompted some answers that challenge the very basis of science.

Vitalism

Perhaps the most baffling thing about biological organisms is their *teleological* quality (or teleonomic, to use the preferred modern term). As noted in Chapter 1, Aristotle introduced the idea that final causes direct their activity towards a goal. Although final causation is anathema to scientists, the teleological flavour of biological systems is undeniable. This presents the scientist with a disturbing quandary. Thus Monod agonizes:[4]

> Objectivity nevertheless obliges us to recognize the teleonomic character of living organisms, to admit that in their structure and performance they decide on and pursue a purpose. Here therefore, at least in appearance, lies a profound epistemological contradiction.

The mysterious qualities of living organisms are so conspicuous that they have often led to the conclusion that living systems represent a class apart, a form of matter and energy that is *so* strange that it defies the laws that enslave ordinary matter and energy.

The belief that life cannot be explained by ordinary physical laws, and therefore requires some sort of 'extra ingredient', is known as vitalism. Vitalists claim that there is a 'life force' or '*élan vital*' which infuses biological systems and accounts for their extraordinary powers and abilities.

Vitalism was developed in great detail in the early years of this century by the embryologist Hans Dreisch, who revived some old Aristotelian ideas of animism. Dreisch postulated the existence of a causal factor operating in living matter called *entelechy*, after the Greek *telos*, from which the word teleology derives. Entelechy implies that the perfect and complete idea of the organism exists in advance. This is intended to suggest that systems with entelechy are goal-oriented, that is, they contain within themselves a blueprint or plan of action. Entelechy therefore acts as a sort of organizing force that arranges physical and chemical processes within an organism in accordance with this goal. For example, the development of an embryo from an egg is guided by entelechy, which somehow contains the blueprint for the finished individual. Dreisch also hoped that entelechy would explain higher modes of biological activity, such as behaviour and purposive action.

Dreisch published his work at time when physics was strongly deterministic, and his ideas about entelechy came into direct conflict with the laws of mechanics. Somehow entelechy has to induce the molecules of a living system to conform to the global plan, which by hypothesis they are not supposed to be able to accomplish on their own. This means, at rock bottom, that a molecule that would have moved to place A in the absence of entelechy, has to move instead to place B. The question then arises as to the nature of the extra force that acts upon it, and the origin of the energy thereby imparted to the molecule. More seriously, it was not at all clear how blueprint information which is not stored anywhere in space can nevertheless bring about the action of a force at a point in space.

Dreisch tried to explain this molecular action by postulating that entelechy was somehow able to temporarily suspend microphysical processes, and therefore affect the *timing* of events on a very small scale. The cumulative effect of many such microscopic interruptions would then bring about the required global changes.

In spite of the compelling simplicity of vitalist ideas, the theory was always regarded as intellectually muddled and disreputable. Today it is completely disregarded.

Mechanism

In stark contrast to vitalism is the *mechanistic* theory of life. This maintains that living organisms are complex machines which function according to the usual laws of physics, under the action of ordinary physical forces. Differences between animate and inanimate matter are attributed to the different levels of complexity alone. The building blocks of 'organic machines' are biochemical molecules (hence, ultimately, the atoms of which these are composed), and an explanation for life is sought by reducing the functions of living organisms to those of the constituent molecular components.

Almost all modern biologists are mechanists, and the mechanistic paradigm is responsible for remarkable progress in understanding the nature of life. This is chiefly due to the impressive advances made in establishing the details of the molecular basis for life, such as the discovery of the form of many biochemical molecules and the 'cracking of the genetic code'. This has encouraged the belief that all biological processes can be understood by reference to the underlying molecular structure, and by implication, the laws of physics. One hears it said that biology is just a branch of chemistry, which is in turn just a branch of physics.

The mechanistic theory of life makes liberal use of machine jargon. Living cells are described as 'factories' under the ultimate 'control' of DNA molecules, which organize the 'assembly' of basic molecular 'units' into larger structures according to a 'program' encoded in the molecular machinery. There is much discussion of 'templates' and 'switching' and 'error correction'. The basic processes of life are identified with activity entirely at the molecular level, like some sort of microscopic Meccano or Lego set.

Terrestrial life is found to be a delicately arranged cooperative venture between two distinct classes of very large molecules: nucleic acids and proteins. The nucleic acids are usually known by their abbreviations, RNA and DNA. In most organisms it is the DNA which contains the genetic information. DNA molecules, which may contain millions of atoms strung together in a precise double-helix pattern, do very little else. They are the 'master files', storing the blueprints needed for replication. Francis Crick, co-discoverer of the geometrical form of DNA, has a more picturesque description. DNA molecules, he says, are the 'dumb blondes' of molecular biology – well suited to reproduction, but not much use for anything else.

Most of the work at the molecular level is performed by the proteins, which go to make up much of the structure of the organism and also carry out the main housekeeping tasks. Proteins, which may contain thousands of atoms, are formed as long chains of smaller units, called amino acids, with a variety of side chains hanging on. The entire assemblage must then fold itself up into an intricate and rather specific three-dimensional structure before it can function correctly. One of the characteristic features of proteins is that they are all composed of exactly the same set of amino acids, 20 in number. Whether in a bear, begonia or bacterium, the same 20 amino acids are employed.

The structure of DNA is also based on long chains of similar units, with attendant side-groups. The backbone of the molecule consists of an alternating sequence of phosphate and sugar molecules, with just four different types of side-groups, called bases, hanging on to the sugars. These four bases form the letters of the 'genetic code', and are known by their abbreviations, A,G,T and C. The sizes and shapes of the bases are such that A fits together neatly with T, and G with C. In its normal form a DNA molecule consists of two such chains, clinging together at each 'rung' by complementary base-pairs, the whole agglomeration being coiled into a helical shape – the famous double helix. An important feature of this arrangement is that the molecular bonds within each chain are fairly sturdy, whereas the cross links between the chains are rather weak. This enables the pair of chains to be unzipped without destroying the crucial sequence of bases A,G,T,C on each chain. This is the essence of the system's ability to replicate without errors being introduced.

The cooperative relationship between DNA and proteins requires a mechanism for translating the four-letter DNA code into the 20-letter protein code. The dictionary for this translation was discovered in the sixties. The base sequence is read off from the DNA in units of three at a time, each triplet corresponding to a particular amino acid. The way in which proteins are assembled using the information stored on the DNA is rather complicated. Sections of base sequences on the DNA are copied on to single strands of the related RNA molecule, which acts as a messenger. The instructions for building proteins are conveyed by this messenger RNA to protein factories called ribosomes – very complex molecules that make use of yet another form of nucleic acid called tRNA.

The job of transcribing the instructions for protein assembly, translating it from the four-letter nucleic acid language into the 20-letter protein language, and then finally synthesizing the proteins from available components in the form of amino acids, is strongly reminiscent of a computer controlled automobile production line.

The complex network of operations involves a high degree of feedback. It would not work at all were it not for a key property of proteins, that they can act as enzymes – chemical catalysts – which drive the necessary chemical changes by breaking or cementing molecular bonds. Enzymes are rather like the assembly line technicians (or computer-controlled arms) that reach inside the complex machinery to drill a hole or weld a joint at a crucial place.

Can life be reduced to physics?

It is clear from the foregoing that the microscopic components of an organism consist of a society of molecules, each apparently responding blindly to the physical forces that happen to act upon them at that point in space and time, yet somehow cooperating and integrating their individual behaviour into a coherent order. With the marvellous advances of modern molecular biology, we can now see in detail the clash of ideas that dates from that ancient conflict: Democritus' atomism and Aristotle's holistic teleology. How can individual atoms, moving strictly in accordance with the causal laws of physics, responding only to *local* forces that happen to be produced by neighbouring atoms, nevertheless act collectively in a purposeful, organized and cooperative fashion over length scales vastly in excess of intermolecular distances? This is Monod's 'profound epistemological contradiction' referred to above.

In spite of the fierce mechanistic leanings of modern biologists, such a contradiction inevitably surfaces sooner or later if an attempt is made to reduce all biological phenomena to molecular physics. Thus geneticist Giuseppe Montalenti writes:[5]

Structural and functional complexity of organisms, and above all the finalism of biological phenomena, have been the insuperable difficulty, the insoluble aporia preventing the acceptance of a mechanistic interpretation of life. This is the main reason why in the competition of Aristotelian and Democritean interpretations the former has been the winner, from the beginning to our days.

All attempts to establish a mechanistic interpretation were frustrated by the following facts: (a) the inadequacy of physical laws to explain biological finalism; (b) the crudeness of physical schemes for such fine and complex phenomena as the biological ones; (c) the failure of 'reductionism' to realize that at each level of integration occurring in biological systems new qualities arise which need new explanatory principles that are unknown (and unnecessary) in physics.

Much of the debate between biological reductionists and their opponents takes place, however, at cross purposes. Reductionistic biologists take the position that once the basic physical mechanisms operating in a biological organism have been identified, life has been explained as 'nothing but' the processes of ordinary physics. They argue that because each component of a living organism fails to reveal any sign of peculiar forces at work, life has already effectively been reduced to ordinary physics and chemistry. Since animate and inanimate matter experience exactly the same sort of forces, and since many of life's processes can be conducted in a test tube, any outstanding gaps in knowledge are attributed solely to technical limitations. As time goes on, it is claimed, more and more details of the workings of organisms will be understood within the basic mechanistic paradigm.

It is worth pointing out that the claim that animate and inanimate matter are both subject to the same physical forces is very far from being tested in practice. What the biologist means is that he sees no reason why the sort of molecular activity he studies should not be consistent with the operation of normal physical forces, and that should anyone decide to investigate more closely the biologist would not expect any conflict with conventional physics and chemistry to emerge.

Let us nevertheless grant that the biologist may be right on this score. It is still far from the case, however, that life has then been 'explained' by physics. It has, rather, simply been defined away. For if animate and inanimate matter are indistinguishable in their behaviour under the laws of physics then wherein lies the crucial distinction between living and non-living systems? This point has been emphasized by the physicist Howard Pattee, who has had a longstanding interest in the nature of life. He writes:[6] 'We do not find the physical similarity of living and nonliving matter so puzzling as the observable differences.' To argue the latter away 'is to miss the whole problem'.

The mystery of life, then, lies not so much in the nature of the forces that act on the individual molecules that make up an organism, but in how the whole assemblage operates collectively in a coherent and cooperative fashion. Biology will never be reconciled with physics until it is recognized that each new level in the hierarchical organization of matter brings into existence new qualities that are simply irrelevant at the atomistic level. Scientists are coming increasingly to recognize that there is no longer any basis in physics for this sort of reductionism. In Chapter 4 it was explained how non-linear systems can display chaotic, unpredictable behaviour that cannot be analysed into the activity of component subsystems. Writing about chaos in a recent issue of *Scientific American* a group of physicists pointed out that:[7]

> Chaos brings a new challenge to the reductionist view that a system can be understood by breaking it down and studying each piece. This view has been prevalent in science in part because there are so many systems for which the behaviour of the whole is indeed the sum of its parts. Chaos demonstrates, however, that a system can have complicated behaviour that emerges as a consequence of simple, nonlinear interaction of only a few components. The problem is becoming acute in a wide range of scientific disciplines, from describing microscopic physics to modelling macroscopic behaviour of biological organisms . . . For example, even with a complete map of the nervous system of a simple organism . . . the organism's behaviour cannot be deduced. Similarly, the hope that physics could be complete with an increasingly detailed understanding of fundamental physical forces and constituents is unfounded. The interaction of components on one scale can lead to complex global behaviour on a larger scale that in general cannot be deduced from knowledge of the individual components.

Morphogenesis: the mystery of pattern formation

Among the many scientific puzzles posed by living organisms, perhaps the toughest concerns the origin of form. Put simply, the problem is this. How is a disorganized collection of molecules assembled into a coherent whole that constitutes a living organism, with all the right bits in the right places? The creation of biological forms is known as morphogenesis, and despite decades of study it is a subject still shrouded in mystery.

The enigma is at its most striking in the seemingly miraculous development of the embryo from a single fertilized cell into a more or less independent living entity of fantastic complexity, in which many cells have become specialized to form parts of nerves, liver, bone, etc. It is a process that is somehow supervised to an astonishing level of detail and accuracy in both space and time.

In studying the development of the embryo it is hard to resist the impression that there exists somewhere a blueprint, or plan of assembly, carrying the instructions needed to achieve the finished form. In some as yet poorly understood way, the growth of the organism is tightly constrained to conform to this plan. There is thus a strong element of teleology involved. It seems as if the growing organism is being directed towards its final state by some sort of global supervising agency. This sense of destiny has led biologists to use the term *fate map* to describe the seemingly planned unfolding of the developing embryo.

Morphogenesis is all the more remarkable for its robustness. The developing embryos of some species can be mutilated in their early stages without affecting the end product. The ability of embryos to rearrange their growth patterns to accommodate this mutilation is called *regulation*. Regulation can involve new cells replacing removed ones, or cells that have been repositioned finding their way back to their 'correct' locations. It was experiments of this sort that led Driesch to reject any hope of a mechanistic explanation and to opt instead for his vitalist theory.

Although mutilation of the developing organism is often irreversible after a certain stage of cell specialization, there are organisms that can repair damage even in their adult form. Flatworms, for example, when chopped up, develop into several complete worms. Salamanders can regenerate an entire new limb if one is removed. Most bizarre of all is the hydra, a simple creature consisting of a trunk crowned by tentacles. If a fully developed hydra is minced into pieces and left, it will reassemble itself in its entirety!

If there is a blueprint, the information must be stored somewhere, and the obvious place is in the DNA of the original fertilized egg, known to be the repository of genetic information. This implies that the 'plan' is molecular in nature. The problem is then to understand how the spatial arrangement of something many centimetres in size can be organized from the molecular level. Consider, for example, the phenomenon of cell differentiation. How do some cells 'know' they have to become blood cells, while others must become part of the gut, or backbone? Then there is the problem of spatial positioning. How does a given cell know where it is located in relation to other parts of the organism, so that it can 'turn into' the appropriate type of cell for the finished product?

Related to these difficulties is the fact that although different parts of the organism develop differently, they all contain the same DNA. If every molecule of DNA possesses the same global plan for the whole organism, how is it that different cells implement different parts of that plan? Is there, perhaps, a 'metaplan' to tell each cell which part of the plan to implement? If so, where is the metaplan located? In the DNA? But this is surely to fall into an infinite regress.

At present biologists are tackling these puzzles by concentrating their research on the theory of gene switching. The idea is that certain genes within the DNA strand are responsible for certain developmental tasks. Normally these genes lie dormant, but at the appropriate moment they are somehow 'switched on' and begin their regulatory functions. The sequencing of gene switching is therefore most important. When it goes wrong the organism may turn into a monster, with anatomical features appearing in the wrong places. Experiments with fruit flies have produced many such monstrosities. This research has led to the identification of a collection of master genes called the homeobox, which seems to be present in other organisms too, including man. Its ubiquity suggests it plays a key role in controlling other genes that regulate cell differentiation.

Exciting though these advances are, they really concern only the *mechanism* of morphogenesis. They fail to address the deeper mystery of how that mechanism is made to conform with a global plan. The real challenge is to demonstrate how *localized* interactions can exercise *global* control. It is very hard to see how this can ever be explained in mechanistic terms at the molecular level.

What help can we gain from studying other examples of the growth of form in nature?

In the previous sections we have seen how many physical and chemical systems involving local interactions can nevertheless display spontaneous self-organization, producing new and more complex forms and patterns of activity. It is tempting to believe that these processes provide the basis for biological morphogenesis. It is certainly true that, generally speaking, non-linear feedback systems, open to their environment and driven far from equilibrium, will become unstable and undergo spontaneous transitions to states with long-range order, i.e. display global organization.

In the case of the embryo, the initial collection of cells forms a homogeneous mass, but as the embryo develops this spatial symmetry is broken again and again, forming an incredibly intricate pattern. It is possible to imagine that each successive symmetry breaking is a *bifurcation process*, resulting from some sort of chemical instability of the sort discussed in Chapter 6. This approach has been developed in much detail by the French mathematician René Thom using his famous theory of catastrophes. (Catastrophe theory is a branch of topology which addresses discontinuous changes in natural phenomena, and classifies them into distinct types.)

There is, however, a deep problem of principle involved in comparing biological morphogenesis with the growth of structure in simple chemical systems. The global organization in, say, convection cells is of a fundamentally different character from the biological case, because it is spontaneous. It happens in spite of the fact that there is *no* 'global plan' or 'fate map' for these systems. The convection cells do not form according to a blueprint encoded in the fluid molecules. In fact, the convective instability is unpredictable and uncontrollable in its detailed form. Moreover, such control as there may be has to be exercised *through the manipulation of boundary conditions*, i.e. it is irreducibly *global and holistic* in nature.

By contrast, the essential feature of biological organization is that the long-range order of an organism is far from being spontaneous and unpredictable. Given the structure of the DNA, the final form is determined to an astonishing level of detail and accuracy. And whereas a phenomenon such as convective instability is exceedingly sensitive to random microscopic fluctuations, biological morphogenesis is, as we have seen, surprisingly robust.

Somehow the microscopic one-dimensional strand of genetic information has to exercise a coordinating influence, both spatial and temporal, over the *collective* activity of billions of cells spread across what is, size for size, a vast region of three-dimensional space. Identifying physical processes, such as bifurcation instabilities, that allow physical structures to undergo large abrupt changes in form are undoubtedly relevant to the *mechanism* of morphogenesis. However, they leave open the problem of how such changes can be controlled by an arrangement of microscopic particles, especially as this control is of a *non-local* character involving boundary conditions. It is the relationship between the *locally* stored information and the *global*, *holistic* manipulation necessary to produce the relevant patterns which lies at the heart of the 'miracle' of morphogenesis.

In the face of these difficulties, some biologists have questioned whether the traditional mechanistic reductionist approach can ever be successful, based as it is on the particle concept, borrowed from physics. As remarked earlier, physicists no longer regard particles as primary objects anyway. This role is reserved for fields. So far the field concept has made little impact on biology. Nevertheless, the idea that fields of some sort might be at work in morphogenesis is taken seriously. These 'morphogenetic fields' have been variously identified as chemical concentration fields, electric fields or even fields unknown to present physics.

The activity of fields could help explain biological forms because fields, unlike particles, are extended entities. They are thus better suited to accounting for long-range or global features. However, there still remains the central problem of how the genetic information containing the global plan, which supposedly resides in particle form in the DNA, communicates itself to the fields and manages to impose upon them the requisite pattern. In physics, field patterns are imposed by boundary conditions, i.e. global, holistic control.

There is a further problem about the field concept in morphogenesis. As each cell of a given organism contains the same DNA, it is hard to see how the coupling between a field and a DNA molecule differs from one molecule to another, as it must if they are to develop differently. If the fields tell the DNA molecules where they are located in the pattern, and the DNA molecules tell the fields what pattern to adopt, nothing is explained because the argument is circular.

A possible escape is to suppose that somehow the global plan is stored in the fields themselves, and that the DNA acts as a *receiver* rather than a source of genetic information. This radical possibility has been explored in detail by biologist Rupert Sheldrake, whose controversial ideas I shall touch upon at the end of Chapter 11.

A survey of morphogenesis thus reveals an unsatisfactory picture. There seem to be fundamental problems of principle in accounting for biological forms in terms of reductionistic physics. The scientist can clearly see organizing factors at work in, for example, the development of the embryo, but has little or no idea of how these organizing factors relate to known physics.

In many ways the development of the embryo embodies the central mystery of all biology, which is how totally new structures and qualities can emerge in the progression from inanimate to animate. The problem is present in the collective sense in the biosphere as a whole. This brings us to the subjects of evolution and the origin of life.

8
Life: its Origin and Evolution

Darwin's theory

The fundamental mystery of biology is how such a rich variety of organisms, each so well suited to their particular ecological niche, has come to exist. The Bible proclaims that the various species of living things were simply made that way by God.

The discovery of the dimension of time in biology dramatically transformed the conceptual basis of this mystery. The evidence of geology and palaeontology that the forms of living organisms have changed over billions of years revealed the process of evolution – the gradual alteration, differentiation and adaptation of biological species over aeons of time. Today we know that the first living organisms appeared on Earth over three and a half billion years ago, and were, by present standards, extremely simple. Only gradually, over immense periods of time, have progressively more complex organisms evolved from these simple precursors.

The publication of Charles Darwin's *The Origin of Species* in 1859 was a pivotal event in the history of science, comparable with the publication of Newton's *Principia* a century and a half before. Darwin's theory that evolution is driven by random mutation and natural selection was so spectacularly successful that it precipitated the collapse of the last vestiges of Aristotelian teleology. Already banished from the physical sciences, teleological explanations in terms of final causes could now be discarded from the life sciences too.

In the modern era, the tremendous advances in genetics and molecular biology have served only to strengthen support for the essential ideas of Darwin's theory. In particular, it is now possible to understand something of the mechanism of evolutionary change at the molecular level. Mutations occur when genes, which are groups of molecules that can be studied directly, become rearranged within an organism's DNA. The prevailing view is that these rearrangements are primarily brought about spontaneously by transpositions of genetic elements and accidental copying errors during reproduction.

In spite of its evident success, there have always been dissenters to Darwin's theory, and its modern formulation known as neo-Darwinism. Even today there are distinguished scientists who find the basis of neo-Darwinism implausible. These scientists do not doubt the *fact* of evolution – for the fossil record leaves no room for doubt – but they question the adequacy of the Darwinian *mechanism*, i.e. random mutation and natural selection.

Natural selection is the process whereby, in the continual struggle for resources, badly adapted mutants compete poorly and tend to die. Thus organisms which are better suited to their environment are more likely to survive and reproduce than their less well-adapted competitors. This is hardly a statement to be challenged. Indeed, it is essentially tautological. ('Those organisms better suited to survive will survive better.')

More problematic is the claim that evolutionary change is driven by *random* mutations. To place pure chance at the centre of the awesome edifice of biology is for many scientists too much to swallow. (Even Darwin himself expressed misgivings.) Here are some of the objections raised:

How can an incredibly complex organism, so harmoniously organized into an integrated functioning unit, perhaps endowed with exceedingly intricate and efficient organs such as eyes and ears, be the product of a series of pure accidents?

How can random events have successfully maintained biological adaptation over millions of years in the face of changing conditions?

How can chance alone be responsible for the emergence of *completely new and successful* structures, such as nervous system, brain, eye, etc. in response to environmental challenge?

At the heart of these misgivings lies the nature of random processes and the laws of probability. One does not have to be a mathematician to appreciate that the more intricately and delicately a complex system is arranged the more vulnerable it is to degradation by random changes. A minor error in copying the blueprint of a bicycle, for example, would probably make little difference in the performance of the assembled machine. But even a tiny error in the blueprint of an aircraft or spacecraft might well lead to disaster. The same point can be made with the help of the card-shuffling analogy we have already used in Chapter 2. A highly ordered sequence of cards will almost certainly become *less* ordered as a result of shuffling.

In the same way, one would suppose that random mutations in biology would tend to degrade, rather than enhance, the complex and intricate adaptedness of organisms. This is indeed the case, as direct experiment has shown: most mutations are harmful. Yet it is still asserted that random 'gene shuffling' is responsible for the emergence of eyes, ears, brains, and all the other marvellous paraphernalia of living things. How can this be? Intuitively one feels that shuffling can lead only to chaos, not order.

The problem is sometimes cast in the language of information theory. The information necessary to construct an organism is contained in the genes. The more complex and developed the organism is, the greater the quantity of information needed to specify it. Evidently, as evolution has produced organisms of greater and greater sophistication and complexity, the information content of the DNA has steadily risen. Where has this information come from?

Information theorists have demonstrated that 'noise', i.e. random disturbances, has the effect of *reducing* information. (Just think of having a telephone conversation over a noisy line.) This is, in fact, another example of the second law of thermodynamics; information is a form of 'negative entropy', and as entropy goes up in accordance with the second law, so information goes down. Again, one is led to the conclusion that randomness cannot be a consistent source of order.

Dissent

For some scientists and philosophers the above considerations have suggested that chance alone is hopelessly inadequate to account for the richness of the biosphere. They postulate the existence of some additional organizing forces or guiding principles that drive evolutionary change in the direction of better adaptation and more developed levels of organization. This was, of course, the basis of Aristotle's animism, whereby evolution is directed towards a specific goal by the action of final causes. It is also an extension of the idea of vitalism. Thus the French vitalist philosopher Henri Bergson postulated that his so-called *élan vital*, which supposedly endows living matter with special organizing capabilities, is also responsible for directing evolutionary change in a creative and felicitous manner.

Similar concepts underlie many religious beliefs about evolution. For example, earlier in this century Lecomptes du Nöys argued that evolution does not proceed at random, but is directed towards a pre-established goal by a transcendent deity. The Jesuit palaeontologist Teilhard de Chardin took a rather different position. He proposed, not that evolution is directed in its details according to a pre-existing plan, but that it is shaped overall to converge on a yet-to-be-achieved superior final stage, which he called the 'Omega point', representing communion with God.

In more recent times, the cosmologist and astrophysicist Fred Hoyle and his collaborator Chandra Wickramasinghe take a sort of middle path. They reject chance as a creative force in evolution and theorize instead that evolutionary change is driven by the continual input of genetic material from outside the Earth, in the form of micro-organisms that can survive in interstellar space. In his wide-ranging book *The Intelligent Universe*, Hoyle writes of 'evolution by cosmic control':[1]

> The presence of microorganisms in space and on other planets, and their ability to survive a journey through the Earth's atmosphere, all point to one conclusion. They make it highly likely that the genetic material of our cells, the DNA double helix, is an accumulation of genes that arrived on the Earth from outside.

Going on to discuss the role of mind and intelligence in this context, Hoyle concludes that the crucial genetic bombardment is ultimately under the guidance of a super-intellect operating *within* the physical universe and manipulating our physical, as well as biological, cosmic environment.

These are, of course, examples of extreme dissent from the Darwinian paradigm. There are many other instances from within the scientific community of less sweeping, but nevertheless genuine dissatisfaction with conventional neo-Darwinian theory. Some scientists remain sceptical that random mutation plus natural selection is enough and claim that biological evolution requires additional organizing principles if the existence of the plethora of complex organisms on Earth is to be satisfactorily explained.

What defence can be given against the criticism that chance mutations cannot generate the wonders of biology? The standard response to these misgivings is to point out that random changes will certainly *occasionally* produce improvements in the performance of an organism, and that these improvements will be selectively preserved, distilled and enhanced by the filter of natural selection until they come to predominate.

It is easy to imagine examples. Suppose a group of animals is trapped on an island where changing climate is bringing greater aridity, and suppose it so happens that a random mutation produces an animal that can survive for longer periods without water. Clearly there is a good chance of that particular animal living longer and producing more offspring. The many offspring will inherit the useful trait and go on to propagate it themselves. Thus, gradually, the less-thirsty strain will come to predominate.

Although selective filtering and enhancement of useful genes could obviously occur in the manner just described for individual cases, the explanation has the flavour of a just-so story. It is far more difficult to demonstrate that there will be a *systematic* accumulation of myriads of such changes to produce a coherent pattern of species advancement. Is it adequate to explain the appearance of a 'grand strategy', whereby life on our planet seems to become progressively more successful in exploiting environmental opportunities?

Consider, for example, intricate organs such as the eye and ear. The component parts of these organs are so specifically interdependent it is hard to believe that they have arisen separately and gradually by a sequence of independent accidents. After all, half an eye would be of dubious selective advantage; it would, in fact, be utterly useless. But what are the chances that just the right sequence of purely random mutations would occur in the limited time available so that the end product happened to be a successfully functioning eye?

Unfortunately it is precisely on this key issue that neo-Darwinism necessarily gets vague. Laboratory studies give some idea of the rate at which mutations occur in some species, such as *Drosophila*, and estimates can be made of the ratio of useful to harmful mutations as judged by human criteria of useful. The problem is that there is no way to quantify the selective advantage of mutations in general. How can one know by *how much* a longer tail or more teeth confers advantage in such-and-such an environment. Just *how many* extra offspring do these differences lead to? And even if these answers were known, we cannot know what were the precise environmental conditions and changes that occurred over billions of years, nor the circumstances of the organisms extant at the time.

A further difficulty is that it is not only the environment which is responsible for selection. One must also take into account the behaviour and habits of organisms themselves, i.e. their teleonomic nature, over the aeons. But this 'quality of life' is something that we can know almost nothing about. In short, in the absence of being able to quantify the quality of life, it is hard to see how the adequacy of random mutations can ever be fully tested.

The problem of the arrow of time

The above difficulties are thrown into sharper relief when one considers the evolution of the biosphere as a whole. The history of life has often been described as a *progression* from 'lower' to 'higher' organisms, with man at the pinnacle of biological 'success', emerging only after billions of years of ascent up the evolutionary ladder. Although many biologists dismiss the 'ladder' concept as anthropocentric, it is much harder to deny that in some objective sense life on Earth has at least gradually become more and more complex. Indeed, the tendency for life to evolve from simple to complex is the most explicit example of the general law that organizational complexity tends to increase with time.

It is far from clear how this tendency towards higher levels of organization follows from Darwin's theory. Single celled organisms, for example, are extremely successful. They have been around for billions of years. In their competition with higher organisms, including man, they all too often come out on top, as the medical profession is well aware. What mechanism has driven evolution to produce multicelled organisms of steadily increasing complexity? Elephants may be more interesting than bacteria, but in the strict biological sense are they obviously more successful? In the neo-Darwinian theory success is measured solely by the number of offspring, so it seems that bacteria are vastly more successful than elephants. Why, then, have animals as complex as elephants ever evolved? Why aren't all organisms merely bags of furiously reproducing chemicals? True, biologists can sometimes demonstrate the reproductive advantage of a particular complex organ, but there is no obvious systematic trend apparent.

The evolutionist John Maynard-Smith concedes that the steady accumulation of complexity in the biosphere presents a major difficulty for neo-Darwinism:[2]

Thus there is nothing in neo-Darwinism which enables us to predict a long-term increase in complexity. All one can say is that since the first living organisms were presumably very simple, then if any large change has occurred in evolutionary lineage, it must have been in the direction of increasing complexity; as Thomas Hood might have said, 'Nowhere to go but up' . . . But this is intuition, not reason.

There is, in fact, a deep obstacle of principle to a neo-Darwinian explanation for the progressive nature of evolutionary change. The point about increasing biological complexity is that it is time-asymmetric; it defines an arrow of time from past to future. Any successful theory of evolution must explain the origin of this arrow.

In Chapter 2 we saw how, since the work of Boltzmann, physicists have appreciated that microscopic random shuffling does not alone possess the power to generate an arrow of time, because of the underlying time symmetry of the microscopic laws of motion. On its own, random shuffling merely produces what might be called stochastic drift with no coherent directionality. (The biological significance of this has recently been recognized by the Japanese biologist Kimura who has coined the phrase 'neutral evolution' to describe such directionless drift.[3])

If an arrow of time exists, it comes not from within the system itself, but from *outside*. This can occur in one of two ways. The first way is if a system is created by its environment in a state which initially has less than maximum entropy, and is then closed off as an independent branch system. Under these circumstances, steady descent into chaos follows, as the entropy rises in accordance with the second law of thermodynamics.

Now this is clearly the opposite of what is happening in biology. That does not, of course, mean that biological organisms violate the second law. Biosystems are not closed systems. They are characterized by their very openness, which enables them to export entropy into their environment to prevent degeneration. But the fact that they are able to evade the degenerative (pessimistic) arrow of time does not explain how they comply with the progressive (optimistic) arrow. Freeing a system from the strictures of one law does not prove that it follows another.

Many biologists make this mistake. They assume because they have discovered the above loophole in the second law, the progressive nature of biological evolution is explained. This is simply incorrect. It also confuses order with organization and complexity. Preventing a decrease in *order* might be a necessary condition for the growth of organization and complexity, but it is not a sufficient condition. We still have to find that elusive arrow of time.

Let us therefore turn to the second way of introducing an arrow of time into a physical system. This comes when open systems are driven far from equilibrium. As we have seen from many examples in physics and chemistry, such systems may reach critical 'bifurcation points' at which they leap abruptly into new states of greater organizational complexity. It seems clear that it is this tendency, rather than random mutation and natural selection, that is the essential mode of progressive biological evolution. The concept of randomness is only appropriate when we can apply the usual statistical assumptions, such as the 'law of large numbers'. This law fails at the bifurcation points, where a single fluctuation can become amplified and stabilized, altering the system dramatically and suddenly.

The power behind evolutionary change, then, is the continual forcing of the biosphere away from its usual state of dynamic equilibrium, either by internal or external changes. These can be gradual, such as the slow build-up of oxygen in the atmosphere and the increase in the sun's luminosity, or sudden, as with the impact of an asteroid, or some other catastrophic event. Whatever the reason, if self-organization in biological evolution follows the same general principles as non-biological self-organization we would expect evolutionary change to occur in sudden jumps, after the fashion of the abrupt changes at certain critical points in physical and chemical systems. There is in fact some evidence that evolution has occurred this way.[4]

What are we to conclude from this? Complex structures in biology are unlikely to have come about as a result of purely random accidents, a mechanism which fails completely to explain the evolutionary arrow of time. Far more likely, it seems, is that complexity in biology has arisen as part of the same general principle that governs the appearance of complexity in physics and chemistry, namely the very *non-random* abrupt transitions to new states of greater organizational complexity that occur when systems are forced away from equilibrium and encounter 'critical points'.

It is not necessary to add any mystical or transcendent influences here. There is no reason to suppose that the principles which generate new levels of organization in biology are any more mystical than those that produce the spiral shapes of the Belousov–Zhabatinski reaction. But it is essential to realize that these principles are inherently global, or holistic, and cannot be reduced to the behaviour of individual molecules, although they are compatible with the behaviour of those molecules. Hence my contention that *purely* molecular mechanical explanations of evolution will prove to be inadequate. If non-biological self-organization is anything to go by, we have to look for holistic principles that govern the collective activity of all the components of the organism.

It is interesting that some theoretical biologists have come to similar conclusions from work in automata theory. Stuart Kauffman of the Department of Biochemistry and Biophysics at the University of Pennsylvania has made a study of the behaviour of randomly assembled ensembles of cellular automata and discovered that they can display a wide range of emergent properties that he believes will help explain biological evolution, and even 'hints that hitherto unexpected principles of order may be found'. The automaton rules are generally not time reversible, and are capable of precisely the sort of progressive self-organizing behaviour that occurs in biological evolution. Kauffman concludes that it is this self-organizing behaviour rather than selection that is responsible for evolution:[5]

> A fundamental implication for biological evolution itself may be that selection is not powerful enough to *avoid* the generic self-organized properties of complex regulatory systems persistently 'scrambled' by mutation. Those generic properties would emerge as biological universals, widespread in organisms not by virtue of selection or shared descent, but by virtue of their membership in a common ensemble of regulatory systems.

Origins

The problems concerning the emergence of complexity through evolution pale beside the formidable difficulty presented by the origin of life. The emergence of living matter from non-living matter is probably the most important example of the self-organizing capabilities of physical systems. Given a living organism, it is possible to imagine ways in which it may multiply. But where did the first organism itself come from? Life begets life, but how does non-life beget life?

It should be stated at the outset that the origin of life remains a deep mystery. There are no lack of theories, of course, but the divergence of opinion among scientists on this topic is probably greater than for any other topic in biology.

The essential problem in explaining how life arose is that even the simplest living things are stupendously complex. The replicative machinery of life is based on the DNA molecule, which is itself as structurally complicated and intricately arranged as an automobile assembly line. If replication requires such a high threshold of complexity in the first place how can any replicative system have arisen spontaneously?

The problem is actually understated when put this way. As we have seen, all life involves cooperation between nucleic acids and proteins. Nucleic acids carry the genetic information, but they cannot on their own do anything. They are chemically incompetent. The actual work is carried out by the proteins with their remarkable catalytic abilities. But the proteins are themselves assembled according to instructions carried by the nucleic acids. It is the original chicken and egg problem. Even if a physical mechanism were discovered that could somehow assemble a DNA molecule, it would be useless unless another mechanism simultaneously surrounded it with relevant proteins. Yet it is hard to conceive that the complete interlocking system was produced spontaneously in a single step.

Scientists attempting to solve this riddle have divided into two camps. In the first camp are those who believe that life originated when a chemical structure appeared that could play the role of a gene, i.e. one capable of replication and genetic information storage. This need not have been DNA; in fact, some scientists favour RNA for this honour. It could be that DNA appeared only much later on in evolutionary history. Whatever it was, this primeval genetic chemical had to have arisen and become capable of performing its replicative function without the assistance of protein enzymes to act as catalysts. In the second camp are those who believe that the chemically much simpler proteins arose first, and that genetic capability appeared gradually, through a long period of chemical evolution, culminating in the production of DNA.

Advocates of the nucleic-acid-first group, such as Leslie Orgel of the Salk Institute in La Jolla, California, have tried to induce RNA replication in a test tube without protein assistance. Manfred Eigen, a Nobel prizewinner who works at the Max Planck Institute in Göttingen, Germany, has constructed an elaborate scenario for the origin of life based on experiments with viral RNA (viruses are the simplest living objects known), and the use of complicated mathematical models. He proposes that RNA can form spontaneously from other complex organic chemicals through a hierarchy of interlocking, mutually reinforcing chemical cycles which he refers to as *hypercycles*. These cycles involved some proteins too.

A proponent of the proteins-first school is Sidney Fox of the University of Miami. He has carried out experiments in which assortments of amino acids (important building blocks of organic molecules) are heated to form 'proteinoids' – molecules that resemble proteins. Although proteinoids are not found in living organisms, they exhibit some startlingly lifelike qualities. Most striking is the way that they can form minute spheres that resemble in some respects living cells. This could be taken as a hint that the cellular structure of living organisms came first, with the nucleic acid control evolving afterwards.

A completely different route to life has been proposed by Graham Cairns-Smith of the University of Glasgow, who believes that the first life forms might not have used carbon-based organic compounds at all, but clay. Some clay crystals can perform a rudimentary form of replication, and could, perhaps, provide sufficient complexity for genetic storage and transmission after a fashion. At any rate, the theory is that the primitive clay organisms gradually evolved more complex practices, including experimentation with organic substances. In the fullness of time, the organic molecules took over the genetic function, and the clay origins of life were lost.

All these speculations are a far cry from an actual demonstration that life can arise spontaneously by ordinary chemical processes of the sort that might have taken place naturally on Earth billions of years ago. It has to be conceded that although all the currently popular scenarios *could* have produced life, none of them is compelling enough for us to believe it *had* to have happened that way.

At this point, a word must be said about the famous experiment performed by Stanley Miller and Harold Urey at the University of Chicago in 1952. The experiment was a crude attempt to simulate the conditions that may have prevailed on the Earth three or four billion years ago, at the time that the first living organisms appeared. At that time there was no free oxygen on the Earth. The atmosphere was chemically of a reducing nature. Nobody is sure of the precise composition, even today. Miller and Urey took a mixture of hydrogen, methane and ammonia gases (all common substances in the solar system), together with boiling water, and passed an electric discharge through the mixture, intended as a substitute for lightning. At the end of a week a red-brown liquid had accumulated. The electric spark was switched off, and the liquid analysed. It was found to contain a number of well-known organic compounds important to life, including some amino acids.

Although the products were trivial in relation to the awesome complexity of molecules such as DNA, the results of the experiment had a profound psychological effect. It became possible to envisage a huge natural Miller–Urey experiment taking place on the Earth's primeval surface over millions of years, with the organic products forming an ever-richer soup in the oceans and in warm pools of water on the land.

Given all that time, it was reasoned, more and more complex organic molecules would be formed by the continual chemical reprocessing of the soup's diverse contents, until at last a single sufficiently complicated replicator molecule would have formed. Once this occurred, the replicator would then rapidly multiply, using for raw materials the chemically-rich broth in which it found itself immersed.

It is possible to perform rough calculations of the probability that the endless breakup and reforming of the soup's complex molecules would lead to a small virus after a billion years. Such are the enormous number of different possible chemical combinations that the odds work out at over $10^{2\,000\,000}$ to one against. This mind-numbing number is more than the chances against flipping heads on a coin six million times in a row. Changing from a virus to some hypothetical simpler replicator could improve the odds considerably, but with numbers like this it doesn't change the conclusion: the spontaneous generation of life by random molecular shuffling is a ludicrously improbable event.

Recipe for a miracle

Betting-odds calculations for the spontaneous generation of life by chance have elicited a number of different responses from scientists. Some have simply shrugged and proclaimed that the origin of life was clearly a unique event. This is, of course, not a very satisfactory position, because where a unique event is concerned the distinction between a natural and a miraculous process evaporates. Science can never be said to have explained such an event.

However, it must not be overlooked that the origin of life differs from other events in a crucial respect: it cannot be separated from our own existence. We are here. Some set of events, however unlikely, must have led up to that fact. Had those events not occurred, we should not be here to comment on it. Of course, if we ever obtain evidence that life has formed independently elsewhere in the universe then this point will become irrelevant.

A quite different response has been to conclude that life did not form on Earth at all but came to Earth from elsewhere in the universe, perhaps in the form of micro-organisms propelled through outer space. This hypothesis was advanced by the Swedish Nobel prizewinner Svante Arrhenius many years ago, but has recently been resurrected by Francis Crick, and in a somewhat different form by Fred Hoyle and Chandra

Wickramasinghe. The problem is that it only pushes the riddle back one step. It is still necessary to explain how life formed elsewhere, presumably under different conditions.

A third response is to reject the entire basis of the 'shuffling accident' scenario – the assumption that the chemical processes that led to life were of a random nature. If the particular chemical combinations that proved important in the formation of life were in some way favoured over the others, then the contents of the primeval soup (or whatever other medium one cares to assume) might have been rather rapidly directed along a pathway of increasing complexity, ultimately to primitive self-replicating molecules.

We have seen how the concept of random shuffling belongs to equilibrium thermodynamics. The sort of conditions under which life is believed to have emerged were far from equilibrium, however, and under these circumstances highly non-random behaviour is expected. Quite generally, matter and energy in far-from-equilibrium open systems have a propensity to seek out higher and higher levels of organization and complexity.

The dramatic contrast between the efficiencies of equilibrium and non-equilibrium mechanisms for the production of life from non-life has been emphasized by Jantsch.[6] On the one hand there is the possibility of 'dull and highly unlikely accidents . . . in the slow rhythm of geophysical oscillations and chemical catalytic processes.' On the other hand:

for every conceivable slowly acting random mechanism in an equilibrium world, there is a mechanism of highly accelerated and intensified processes in a non-equilibrium world which facilitates the formation of dissipative structures and thereby the self-organization of the microscopic world.

Prigogine's work on dissipative structures and Eigen's mathematical analysis of hypercycles both indicate that the primeval soup could have undergone successive leaps of self-organization along a very narrow pathway of chemical development. Our present understanding of chemical self-organization is still very fragmentary. It could perhaps be that there are as yet unknown organizing principles operating in prebiotic chemistry that greatly enhance the formation of complex organic molecules relevant to life.

It is an interesting point of history that the communist doctrine of dialectical materialism holds that new laws of organization come into operation as matter reaches higher levels of development. Thus there are biological laws, social laws, etc. These laws are intended to ensure the onward progression of matter towards states of ever-greater organization. The Russian chemist Alexander Oparin was one of the central figures in the development of the modern paradigm of the origin of life, and he strongly subscribed to the theory that life will be the inevitable outcome of self-organizing chemical processes, though whether for reasons of scientific conviction or political expediency is a matter for debate. Unfortunately this same politically motivated doctrine was grotesquely misapplied by the infamous Trofim Lysenko in an attempt to discredit modern genetics and molecular biology. No doubt this episode has prejudiced some biologists against the *scientific* idea of the origin of life as a culmination of progressive chemical development.

A review of current thinking on the origin of life problem thus reveals a highly unsatisfactory state of affairs. It is straining credulity to suppose that the uniquely complex and specific nucleic acid–protein system formed spontaneously in a single step, yet the only generally accepted organizing principle in biology – natural selection – cannot operate until life of some sort gets going. This means either finding some more primitive chemical system that can undergo progressive evolution by natural selection, or else recognizing the existence of non-random organizing principles in chemical development.

9

The Unfolding Universe

Cosmic organization

It is curious that even on the largest scale of size, matter and energy are arranged in a highly non-random way. A casual glance at the night sky, however, reveals little in the way of order. Stars are scattered, it seems, more or less haphazardly.

A small telescope reveals some structure. Here and there, stars are clustered into groups, occasionally forming tight aggregations as many as one million strong. Surveys with more powerful instruments show that the stars in our 'local' region of space are organized into a vast wheel-shaped system called the Galaxy, containing about one hundred billion stars and measuring one hundred thousand light years in diameter. The Galaxy has a distinctive structure, with a crowded central nucleus surrounded by spiral-shaped arms which contain gas and dust as well as slowly orbiting stars. All this is embedded in a large, more or less spherical halo of material which is largely invisible and is also unidentified.

The organization of the Galaxy is not apparent at first sight because we are viewing it from within, but its form is similar to that of many other galaxies that are revealed by large telescopes. Astronomers have classified galaxies into a number of distinct types – spirals, ellipticals, etc. How they got their shapes is still a mystery. In fact, astronomers have only the vaguest ideas of how galaxies formed.

The general principle that induces cosmological material to aggregate is well enough understood. If, among the primeval gases, there existed some inhomogeneity, each overdense region would act as a focus for gravitational attraction, and start to pull in surrounding material. As more and more material was accreted, so discrete entities formed, separated by empty space. Further gravitational contraction, followed by fragmentation, somehow produced the stars and star clusters. Just how the inhomogeneity got there in the first place remains unknown.

Surprisingly, galaxies are not the largest structures in the universe in spite of their immense size. Most galaxies are aggregated into clusters, and even clusters of clusters. There are also huge voids, strings and sheets of galaxies. Again, the origin of this very large-scale structure is ill-understood.

There is much more to the universe than galaxies. Space is full of unseen matter, perhaps in the form of exotic subatomic particles that interact only very weakly with ordinary matter and therefore go unnoticed. Nobody knows what sort of stuff it is. Although it is invisible, the mystery matter produces important gravitational effects. It can, for example, affect the rate at which the universe as a whole expands. It also affects the precise way in which gravitational aggregation takes place, and therefore exerts an influence on the large-scale structure of the universe.

Passing to still larger length scales, it is found that the tendency for matter to aggregate dies away, and the clusters of galaxies are distributed uniformly in space. The best probe of the very large-scale structure of the universe is the background heat radiation generated in the big bang. It bathes the entire universe and has travelled more or less freely almost since the creation. It would therefore bear the imprint of any major departure from uniformity encountered during its multi-billion-year journey across space. By accurately measuring the smoothness of the background heat radiation coming from different directions astronomers have put limits on the large-scale smoothness of the universe to one part in ten thousand.

Cosmologists have long supposed that the universe is uniform in the large, an assumption known as the cosmological principle. The reason for the uniformity is, however, a profound mystery. To investigate it further we turn to the subject of the big bang itself.

The first one second

It is now generally accepted that the universe came into existence abruptly in a gigantic explosion. Evidence for this 'big bang' comes from the fact that the universe is still expanding; every cluster of galaxies is flying apart from every other. Extrapolating this expansion backwards in time indicates that sometime between 10 and 20 billion years ago the entire contents of the cosmos we see today were compressed into a minute volume of space. Cosmologists believe that the big bang represents not just the appearance of matter and energy in a pre-existing void, but the creation of space and time too. The universe was not created *in* space and time; space and time are *part of* the created universe.

On general grounds it is to be expected that the early stages of the explosion were characterized by very rapid expansion and extreme heat. This expectation was confirmed in 1965 with the discovery that the universe is filled everywhere with a uniform bath of heat radiation. The temperature of this cosmic background is about three degrees above absolute zero – a faded remnant of the once fierce primeval fire.

Again by extrapolating backwards in time, it is clear that the state of the universe during the first few seconds must have been one of extreme simplicity, since the temperature was too high for any complex structures, including atomic nuclei, to have existed. Cosmologists suppose that the cosmological material at the dawn of time consisted of a uniform mixture of dissociated subatomic particles in thermodynamic equilibrium.

A test of this assumption is to model the fate of the particle 'soup' as the temperature fell. Below about a billion degrees, the temperature was no longer too great to prevent the fusion of neutrons and protons into complex nuclei. Calculations indicate that during the first few minutes about 25 per cent of the nuclear material would have formed into nuclei of the element helium, with a little deuterium and lithium, and negligible quantities of anything else. The remaining 75 per cent would have been left unprocessed in the form of individual protons, destined to become hydrogen atoms. The fact that astronomers observe the chemical composition of the universe to be about 25 per cent helium and 75 per cent hydrogen provides welcome confirmation that the basic idea of a hot big bang origin for the universe is correct.

In the original version of the big bang theory, which became popular in the 1960s, the universe was considered to have started out with essentially infinite temperature, density and rate of expansion, and to have been cooling and decelerating ever since. The bang itself was placed beyond the scope of science, as were the contents of the 'soup' which emerged from the explosion, and its distribution in space. All these things had simply to be assumed as given, either God-given, or due to very special initial conditions which the scientist did not regard as his job to explain.

Then, during the 1970s, early-universe cosmology received a major stimulus from an unexpected direction. At that time a torrent of challenging new ideas began to flow from high energy particle physics which found natural application to the very early epochs of the big bang. Particle accelerators came into commission that could directly simulate the searing heat of the primeval universe as far back as one trillionth of a second after the initial event, an epoch at which the temperature was many trillions of degrees. In addition, theorists began to speculate freely about physics at energies greatly in excess of this, corresponding to cosmic epochs as early as a 10^{-36} of a second – the very threshold of creation.

This pleasing confluence of the physics of the very large (cosmology) and very small (particle physics) opened up the possibility of explaining many of the distinctive features of the big bang in terms of physical processes in the very early moments, rather than as the result of special initial conditions. For example, there is some evidence that the primordial irregularities in the distribution of matter necessary for galaxies and galactic clusters to grow might be attributable to quantum fluctuations that occurred at around 10^{-32} seconds.

I do not wish to review these exciting developments in depth here, because I have discussed them in my book *Superforce*. However, I should like to bring out a general point relevant to the present theme. In particle physics the key parameter is energy, and the history of particle physics is largely the quest for greater and greater energies at which to collide subnuclear particles. As the energies of experiments (and of theoretical modelling) have been progressively elevated over the years, a trend has become apparent. Generally speaking, the higher the energy, the less structure and differentiation there is both in subatomic matter itself and the forces that act upon it.

Consider, for example, the various forces of nature. At low energy there seem to be four distinct fundamental forces: gravitation and electromagnetism, familiar in daily life, and two nuclear forces called weak and strong. Imagine for the sake of illustration that we could raise the temperature in a volume of space without limit, and thus simulate earlier and earlier epochs of the primeval universe. According to present theories, at a temperature of about 10^{15} degrees (about the current limit for direct experimentation) the electromagnetic force and the weak nuclear force merge in identity. Above this temperature there are no longer four forces, but three.

Theory suggests that with additional elevation of the temperature further amalgamations would take place. At 10^{27} degrees the strong force would merge with the electromagnetic-weak force. At 10^{32} degrees gravitation would join in, producing a single, unified superforce.

The identity of matter undergoes a similar fade-out as the temperature is raised. This is already familiar in ordinary experience. The most structured and distinctive forms of matter are solids. At higher temperatures solids become liquids, then gases, each phase representing a trend towards featurelessness. Additional heating converts a gas into a plasma, in which even the atoms lose their structure and become dissociated into electrons and ions.

At higher temperatures the nuclei of the atoms break up. This was the state of the cosmological material at about one second. It consisted of a uniform mixture of protons, neutrons and electrons. At earlier moments, before about 10^{-6} seconds, the temperature and density of the nuclear particles (protons and neutrons) was so high that their individual identities were lost, and the cosmological material was reduced to a soup of quarks – the elementary building blocks of all subnuclear matter. At this time, therefore, the universe was filled with a simple mixture of various subatomic particles, including a number of distinct species of quarks, electrons, muons, neutrinos and photons.

With further elevation of the temperature, corresponding to still earlier epochs of the universe, the distinguishing properties of these particles begin to evaporate. For example, some particles lose their masses and, along with the photons, move at the speed of light. At ultra-high temperatures, even the distinction between quarks and leptons (the relatively weakly interacting particles such as electrons and neutrinos) becomes blurred.

If some very recent ideas are to be believed, as the temperature reaches the so-called Planck value of 10^{32} degrees, all matter is dissociated into its most primitive constituents, which may be simply a sea of identical strings existing in a ten-dimensional spacetime. Moreover, under these extreme conditions, even the distinction between spacetime and matter becomes nebulous.

Whatever the technical details of any particular theory, the trend is that as the temperature is raised, so there is less and less structure, form and distinction among particles and forces. In the extreme high-energy limit, all of physics seems to dissolve away into some primitive abstract substratum. Some theorists have even gone further and suggested that the very laws of physics also dissolve away at ultra-high energies, leaving pure chaos to replace the rule of law. These bizarre changes that are predicted to take place at high energies have led to a remarkable new perspective of nature. The physical world of daily experience is now viewed as some sort of frozen vestige of an underlying physics that unifies all forces and particles into a bland amalgam.

Symmetry, and how to get less of it

In their recent book *The Left Hand of Creation*, astrophysicists John Barrow and Joseph Silk write:[1] 'If paradise is the state of ultimate and perfect symmetry, the history of the "big bang" resembles that of "paradise lost" . . . The result is the varied universe of broken symmetry that now surrounds us.'

To appreciate this rather cryptic statement, it is necessary to have an understanding of the place of symmetry in modern physics. We have already seen how symmetry breaking is a characteristic feature of self-organizing processes in biology, chemistry and laboratory physics. We shall now see how it plays a key role in cosmology too.

Overt symmetry is found in abundance in nature – in the spherical figure of the Sun, the pattern of a snowflake, the geometrical form of the planetary orbits – and in human artefacts. Hidden symmetry, however, plays an even more important role in physics. Indeed, much of our present understanding of the fundamental forces of nature draws heavily on the concept of abstract symmetries that are not obvious on casual inspection.

As already remarked, the relationship between symmetry and structure is an inverse one. The appearance of structure and form usually signals the breaking of some earlier symmetry. This is because symmetry is associated with a lack of features. One example of an object with symmetry is a sphere. It may be rotated through any angle about its centre without changing its appearance. If, however, a spot is painted on the surface, this rotational symmetry is broken because we can tell when the sphere has been reoriented by looking for the spot. The sphere with the spot still retains some symmetry though. It may be rotated without change about an axis passing through the spot and the centre of the sphere. It may also be reflected in a suitably oriented mirror. But if the surface were covered with many spots, these less powerful symmetries would also be lost.

Several examples have been mentioned of spontaneous symmetry breaking accompanying the appearance of new forms of order; in the Belousov–Zhabatinski reaction for example, where an initially featureless solution generates spatial patterns; in morphogenesis, where a homogeneous ball of cells grows into a differentiated embryo; in ferromagnetism, where symmetrically distributed micromagnets align in long-range order.

In particle physics there exist symmetries that have no simple geometrical expression. Nevertheless they are crucial to our understanding of the laws of physics at the subatomic level. A prime example are the so-called 'gauge symmetries' that provide the key to the unification of the forces of nature. Gauge symmetries have to do with the existence of freedom to continuously redefine ('re-gauge') various potentials in the mathematical description of the forces without altering the values of the forces at each point in space and time. As with geometrical symmetries, so with gauge symmetries, they tend to become spontaneously broken at low temperatures and restored at high temperatures.

It is precisely this effect that occurs with the electromagnetic and weak forces. These two forces are distinctly different in properties at ordinary energies. The electromagnetic force is very much stronger, and has an infinite range while the relatively feeble weak force is restricted to the subnuclear domain. But as we have seen, above about 10^{15} degrees the two forces merge in identity. They become comparable in strength and range, representing the appearance of a new symmetry (a gauge symmetry in fact) that was hidden, or broken, at low energies.

As we have seen, theory predicts that all sorts of other abstract symmetries become restored as the temperature is raised still further. One of these is the deep symmetry that exists in the laws of nature between matter and antimatter. Matter, which is a form of energy, can be created in the laboratory, but it is always accompanied by an equivalent quantity of antimatter. The fact that matter and antimatter are always produced symmetrically in the laboratory raises the intriguing question of why the universe consists of almost 100 per cent matter. What happened to the antimatter? Evidently some process in the early stages of the big bang broke the matter–antimatter symmetry and enabled an excess of matter to be produced.

The history of the universe can therefore be seen as a succession of symmetry breaks as the temperature falls. Starting with a bland amalgam, step by step more structure and differentiation occurs. With each step a distinctive new quality 'freezes out'. First a slight excess of matter over antimatter became frozen into the cosmological material. This probably happened very early on, at about 10^{-32} seconds after the initial explosion. Then the quarks coalesced into nuclear particles at about one microsecond. By the end of the first one second most of the remaining antimatter had been annihilated by contact with matter.

The stage was then set for the next phase of action. At around one minute, helium nuclei formed from the fusion of the neutrons with some of the protons. Much later, after almost a million years had elapsed, the nuclei and electrons combined to form atoms.

As the universe continued to cool so the primitive cosmic material aggregated to form stars, star clusters, galaxies and other astronomical structures. The stars went on to generate complex nuclei, and spew them into space, enabling planetary systems to form, and as their material cooled, so it congealed into crystals and molecules of ever-growing complexity. Differentiation and processing brought into existence all the multifarious forms of matter encountered on Earth, from diamonds to DNA. And everywhere we look, matter and energy are engaged in the further refinement, complexification and differentiation of matter.

The ultimate origin of the universe

No account of the creation of the universe is complete without a mention of its ultimate origin. A popular theory at the time of writing is the so-called *inflationary scenario*. According to this theory the universe came into existence essentially devoid of all matter and energy. One version of the theory proposes that spacetime appeared spontaneously from nothing as a result of a quantum fluctuation. Another version holds that time in some sense 'turns into' space near the origin, so that rather than considering the appearance of three-dimensional space at an instant of time, one instead deals with a four-dimensional space. If this space is taken to curve smoothly around to form an unbroken continuum, there is then no real origin at all – what we take to be the beginning of the universe is no more a physical origin than the north pole is the beginning of the Earth's surface.

Whatever the case, the next step was for this essentially quiescent 'blob' of new-born space to swell at a fantastic and accelerating rate until it assumed cosmic proportions, a process that took only 10^{-32} seconds or so. This is the 'inflation' after which the scenario is named. It turned a 'little bang' into the familiar big bang.

During the inflationary phase a great deal of energy was produced, but this energy was invisible – locked up in empty space in quantum form. When inflation came to an end, this enormous quantity of energy was then released in the form of matter and radiation. Thereafter the universe evolved in the way already described.

During the inflationary phase the universe was in a condition of perfect symmetry. It consisted of precisely homogeneous and isotropic empty space. Moreover, because the expansion rate was precisely uniform, one moment of time was indistinguishable from another. In other words, the universe was symmetric under time reversal and time translation. It had 'being' but no 'becoming'. The end of inflation was the first great symmetry break: featureless empty space suddenly became inhabited by myriads of particles, representing a colossal increase in entropy. It was a strongly irreversible step, that imprinted an arrow of time on the universe which survives to this day.

If one subscribes to inflation, or something like it, then it seems that the universe started out with more or less nothing at all, and step by step the complex universe we see today evolved through a sequence of symmetry breaks. Each step is highly irreversible and generates a lot of entropy, but each step is also creative, in the sense that it releases new potentialities and opportunities for the further organization and complexification of matter. No longer is creation regarded as a once-and-for-all affair; it is an ongoing process which is still incomplete.

The self-regulating cosmos

The steady unfolding of cosmic order has led to the formation of complex structures on all scales of size. Astronomically speaking, the smallest structures are to be found in the solar system. It is a curious thought that although the motions of the planets have long provided one of the best examples of the successful application of the laws of physics, there is still no proper understanding of the origin of the solar system.

It seems probable that the planets formed from a nebula of gas and dust that surrounded the Sun soon after its formation about five billion years ago. As yet scientists have only a vague idea of the physical processes that were involved. In addition to gravitation there must have been complex hydrodynamic and electromagnetic effects. It is remarkable that from a featureless cloud of swirling material, the present orderly arrangement of planets emerged. It is equally remarkable that the regimented motion of the planets has remained stable for billions of years, in spite of the complicated pattern of mutual gravitational forces acting between the planets.

The planetary orbits possess an unusual, even mysterious, degree of order. Take, for example, the famous Bode's law (actually due to the astronomer Titius) which concerns the distances of the planets from the Sun. It turns out that the simple formula $r_n = 0.4 + 0.3 \times 2^n$, where r_n is the orbital radius of planet number n from the Sun measured in units of the

Earth's orbital radius, fits to within a few per cent all the planets except Neptune and Pluto. Bode's law was able to correctly predict the existence of the planet Uranus, and even predicts the presence of a 'missing' planet where the asteroid belt is located. In spite of this success, there is no agreed theoretical basis for the law. Either the orderly arrangement of the planets is a coincidence, or some as yet unknown physical mechanism has operated to organize the solar system in this way.

Several of the outer planets possess miniature 'solar systems' of their own, in the form of multiple moons and, more spectacularly, rings. The rings of Saturn, to take the best-known example, have aroused the fascination and puzzlement of astronomers ever since their discovery by Galileo in 1610. Forming a huge planar sheet hundreds of thousands of kilometres in size, they give the superficial impression of a continuous solid, but, as remarked in Chapter 5, the rings are really composed of myriads of small orbiting particles.

Close-up photography by space probes has revealed an astonishing range of features and structures that had never been imagined to exist. The apparently smooth ring system was revealed as an intricately complex superposition of thousands of rings, or ringlets, separated by gaps. Less regular features were found too, such as 'spokes', kinks and twists. In addition, many new moonlets, or ringmoons, were discovered embedded in the ring system.

Attempts to build a theoretical understanding of Saturn's rings have to take into account the gravitational forces on the ring particles of the many moons and moonlets of Saturn, as well as the planet itself. Electromagnetic effects as well as gravity play a part. This makes for a highly complicated non-linear system in which many structures have evidently come about spontaneously, through self-organization and cooperative behaviour among the trillions of particles.

One prominent effect is that the gravitational fields of Saturn's moons tend to set up 'resonances' as they orbit periodically, thereby sweeping the rings clear of particles at certain specific radii. Another effect is caused by the gravitational perturbations of moonlets orbiting within the rings. Known as shepherding, it results in disturbances to ring edges, causing the formation of kinks or braids.

There is no proper theoretical understanding of Saturn's rings. In fact, calculations repeatedly suggest that the rings ought to be unstable and decay after an astronomically short duration. For example, estimates of the transfer of momentum between shepherding satellites and the rings indicates that the ring–ringmoon system should collapse after less than one hundred million years. Yet it is almost certain that the rings are *billions* of years old.

The case of Saturn's rings illustrates a general phenomenon. Complex physical systems have a tendency to discover states with a high degree of organization and cooperative activity which are remarkably stable. The study of thermodynamics might lead one to expect that a system such as Saturn's rings, that contains a vast number of interacting particles, would rapidly descend into chaos, destroying all large-scale structure. Instead, complex patterns manage to remain stable over much longer time scales than those of typical disruptive processes. It is impossible to ponder the existence of these rings without words such as 'regulation' and 'control' coming to mind.

An even more dramatic example of a complex system exercising a seemingly unreasonable degree of self-regulation is the planet Earth. A few years ago James Lovelock introduced the intriguing concept of Gaia. Named after the Greek Earth goddess, Gaia is a way of thinking about our planet as a holistic self-regulating system in which the activities of the biosphere cannot be untangled from the complex processes of geology, climatology and atmospheric physics.

Lovelock contemplated the fact that over geological timescales the presence of life on Earth has profoundly modified the environment in which that same life flourishes. For example, the presence of oxygen in our atmosphere is a direct result of photosynthesis of plants. Conversely, the Earth has also undergone changes which are not of organic origin, such as those due to the shifting of the continents, the impact of large meteors and the gradual increase in the luminosity of the Sun. What intrigued Lovelock is that these two apparently independent categories of change seem to be linked.

Take, for example, the question of the Sun's luminosity. As the Sun burns up its hydrogen fuel, its internal structure gradually alters, which in turn affects how brightly it shines. Over the Earth's history the luminosity has increased about 30 per cent. On the other hand the temperature of the Earth's surface has remained remarkably constant over this time, a fact which is known because of the presence of liquid water throughout; the oceans have neither completely frozen, nor boiled. The very fact that life has survived over the greater part of the Earth's history is itself testimony to the equability of conditions.

Somehow the Earth's temperature has been regulated. A mechanism can be found in the level of carbon dioxide in the atmosphere. Carbon dioxide traps heat, producing a 'greenhouse effect'. The primeval atmosphere contained large quantities of carbon dioxide, which acted as a blanket and kept the Earth warm in the relatively weak sunlight of that era. With the appearance of life, however, the carbon dioxide in the atmosphere began to decline as the carbon was synthesized into living material. In compensation, oxygen was released.

This transformation in the chemical make-up of the Earth's atmosphere was most felicitous because it matched rather precisely the increasing output of heat from the Sun. As the Sun grew hotter, so the carbon dioxide blanket was gradually eaten away by life. Furthermore, the oxygen produced an ozone layer in the upper atmosphere that blocked out the dangerous ultra-violet rays. Hitherto life was restricted to the oceans. With the protection of the ozone layer it was able to flourish in the exposed conditions on land.

The fact that life acted in such a way as to maintain the conditions needed for its own survival and progress is a beautiful example of self-regulation. It has a pleasing teleological quality to it. It is as though life anticipated the threat and acted to forestall it. Of course, one must resist the temptation to suppose that biological processes were guided by final causes in a specific way. Nevertheless, Gaia provides a nice illustration of how a highly complex feedback system can display stable modes of activity in the face of drastic external perturbations. We see once again how individual components and sub-processes are guided by the system as a whole to conform to a coherent pattern of behaviour.

The apparently stable conditions on the surface of our planet serves to illustrate the general point that complex systems have a an unusual ability to organize themselves into stable patterns of activity when *a priori* we would expect disintegration and collapse. Most computer simulations of the Earth's atmosphere predict some sort of runaway disaster, such as global glaciation, the boiling of the oceans, or wholesale incineration due to an overabundance of oxygen setting the world on fire. The impression is gained that the atmosphere is only marginally stable. Yet somehow the integrative effect of many interlocking complex processes has maintained atmospheric stability in the face of large-scale changes and even during periods of cataclysmic disruption.

Gravity: the fountainhead of cosmic order

Of the four fundamental forces of nature only gravitation acts across cosmological distances. In this sense, gravity powers the cosmos. It is responsible for bringing about the large-scale structure of the universe, and it is within this structure that the other forces act out their roles.

It has long been appreciated by physicists and astronomers that gravity is peculiar in the way that it organizes matter. Under the action of gravity, a homogeneous gas is unstable. Minor density perturbations will cause some regions of the gas to pull harder than others, causing the surrounding material to aggregate. This accumulation enhances the perturbations and leads to further heterogeneity which may lead the gas to fragment into separated entities. As the material concentrates into definite regions, so the gravitating power of these regions grows. As we have seen, this escalating process may eventually lead to the formation of galaxies and stars. It may even result in the complete collapse of matter into black holes.

This tendency for gravitation to cause matter to grow more and more clumpy is in contrast with the behaviour of a gas on a small scale, where gravitational forces are negligible. An irregularly distributed gas will then rapidly homogenize, as the chaotic agitation of its molecules 'shuffles' it into a uniform distribution. Normally the laws of thermodynamics bring about the disintegration of structure, but in gravitating systems the reverse happens: structures seem to grow with time.

The 'anti-thermodynamic' behaviour of gravity leads to some oddities. Most hot objects, for example, become cooler if they lose energy. Self-gravitating systems, however, do the opposite: they grow hotter. Imagine, for example, that by some magic we could suddenly remove all the heat energy from the Sun. The Sun would then shrink because its gravity would no longer be balanced by its internal pressure. Eventually a new balance would be struck as the compression of the Sun's gases caused their temperature, and hence pressure, to rise. They would need to rise well beyond the present level in order to counteract the higher gravity produced by the more compact state. The Sun would eventually settle down to a new state with a smaller radius and higher temperature.

A practical example of the same basic effect is observed in the decay of satellite orbits. When a satellite brushes the Earth's atmosphere it loses energy and eventually either burns up or falls to the ground. Curiously, though, as the satellite's energy is sapped due to air friction, it actually moves faster, because gravity pulls it into a lower orbit, causing it to gain speed as it goes. This is in contrast to the effect of air resistance near the Earth's surface, which causes bodies to slow down.

The key to the unique structuring capabilities of gravity is its universally attractive nature and long range. Gravity pulls on every particle of matter in the universe and cannot be screened. Its effects are therefore cumulative and escalate with time. As gravitational force draws matter together its strength grows for two reasons. First, the accumulation of matter enhances the source of the pull. Second, the force of gravity rises as matter is compressed due to the inverse square law.

Gravity may be contrasted with the electromagnetic force which is responsible for the behaviour of most everyday systems. This force is also long ranged, but because of the existence of both positive and negative electric charges, electromagnetic fields tend to be screened. The field of an electric dipole (positive and negative charge side by side) diminishes much more rapidly with distance than that of an isolated charge. In effect, then, electromagnetic forces are short ranged; the so-called van der Waal's forces between molecules, for instance, fall off like the inverse seventh power of the distance. It is for this reason that the existence of long-range order in chemical systems such as the Belousov–Zhabatinski reaction is so surprising. But because gravity can reach out across astronomical distances it can exert long-range ordering directly.

These qualities of gravity imply that all material objects are fundamentally *metastable*. They exist only because other forces operate to counteract gravity. If gravity were nature's only force, all matter would be sucked into regions of accumulation and compressed without limit in the escalating gravitational fields there. Matter would, in effect, disappear. Objects such as galaxies and star clusters exist because their rotational motions counteract their gravity with centrifugal forces. Most stars and planets call upon internal pressure of basically electromagnetic origin. Some collapsed stars require quantum press-ure of an exotic origin to survive.

All these states of suspension are, however, vulnerable. When large stars burn out, they lose the battle against their own gravity, and undergo total collapse to form black holes. In a black hole, matter is crushed to a so-called singularity, where it is annihilated. Black holes may also form at the centres of galaxies or star clusters, when the rotational motions become inadequate to prevent matter accumulating above a critical density. Once formed, these black holes may then swallow up other objects which would otherwise have been able to resist their individual self-gravity.

Cosmologists thus see the history of the universe as matter engaged in one long struggle against gravity. Starting with a relatively smooth distribution of matter, the cosmos gradually grows more and more clumpy and structured, as matter descends first into clusters, then clusters of clusters, and so on, leading in the end to black holes. Without gravity, the universe would have remained a panorama of featureless inert gas.

The pulling together of the primeval gases was the crucial step in the formation of galaxies and stars. Once these had formed, the way lay open for the production of the heavy elements, the planets, the vast range of chemical substances, biology and eventually man. In this sense, then, gravity is the fountainhead of all cosmic organization. Way back in the primeval phase of the universe, gravity triggered a cascade of self-organizing processes – organization begets organization – that led, step by step, to the conscious individuals who now contemplate the history of the cosmos and wonder what it all means.

Gravity and the thermodynamic enigma

The ability of gravity to induce the appearance of structure and organization in the universe seems to run counter to the spirit of the second law of thermodynamics. In fact, the relationship between gravity and thermodynamics is still being clarified. It is certainly possible to generalize thermodynamic concepts such as temperature and entropy to self-gravitating systems, but the thermodynamic properties of these systems remain unclear.

For a time it was believed that black holes actually transcend the second law because of their ability to swallow entropy. In the early 1970s, however, Jacob Bekenstein and Stephen Hawking showed that the concept of entropy can be generalized to include black holes (the entropy of a black hole is proportional to its surface area). The key step here was Hawking's demonstration that black holes are not strictly black after all, but have an associated temperature. In many respects the exchange of energy and entropy between a black hole and its environment complies with the same thermodynamic principles that engineers use for heat engines. As might be expected, though, black holes follow the rule of all self-gravitating systems: they grow *hotter* as they radiate energy. In spite of this, the crucial second law of thermodynamics survives intact, once the black hole's own entropy is taken into account.

Though the tendency for self-gravitating systems to grow more clumpy with time does not, after all, contradict the second law of thermodynamics, it is not explained by it either. Once again there is a missing arrow of time. The unidirectional growth of clumpiness in the universe is so crucial to the structure and evolution of the universe that it seems to have the status of a fundamental principle.

One person who believes there is a deep principle involved is Roger Penrose, whose work on the tiling of the plane was mentioned in Chapter 6. Penrose suggests that there is a cosmic law or principle that requires the universe to begin, crudely speaking, in a smooth condition. He has tentatively proposed an explicit mathematical quantity (the Weyl tensor) as a measure of gravitational irregularity, and tried to show that it can only increase under the action of gravitational evolution. His hope is that this quantity can be taken as a measure of the *entropy* of the gravitational field itself, so that the growth in clumpiness can then be regarded as just another example of the ubiquitous growth of entropy with time, i.e. the second law of thermodynamics. One requirement, of course, is that this expression for gravitational entropy goes over to Hawking's above-mentioned area formula in the limiting case that the clumpiness extremizes itself in the form of black holes. Though Penrose seems to be addressing a real and important property of nature, these attempts to make his ideas more rigorous have not yet been carried through convincingly, and Penrose himself now expresses reservations about them.

I believe, with Penrose, that the structuring tendency of self-gravitating systems is the manifestation of a fundamental principle of nature. In fact, it is merely one aspect of the general principle being expounded in this book that the universe is progressively self-organizing. What I believe is needed here, however, is once again a clear distinction between *order* and organization. A clumpy arrangement of self-gravitating matter does *not*, I submit, have more order than a smooth arrangement, but it *does* have a higher degree of organization – just compare a galaxy, with its spiral arms and coherent motion, with a featureless cloud of primeval gas. I think, therefore, that the self-structuring tendency of gravitating systems will not be explained using the concept of gravitational entropy alone, but will require some quantitative measure of the *quality* of gravitational arrangement.

10

The Source of Creation

A third revolution

There is a widespread feeling among physicists that their subject is poised for a major revolution. As already remarked, true revolutions in science are not just rapid advances in technical details, but transformations of the concepts upon which science is based. In physics, revolutions of this magnitude have occurred twice before. The first was the systematic development of mechanics by Galileo and Newton. The second occurred with the theory of relativity and the quantum theory at the beginning of this century.

On one front, great excitement is being generated by the ambitious theoretical attempts to unify the forces of nature and provide a complete description of all subatomic particles. Such a scheme has been dubbed a 'Theory of Everything', or TOE for short. This programme, which has grown out of high energy particle physics and has now made contact with cosmology, is a search for the ultimate principles that operate at the lowest and simplest level of physical description. If it succeeds, it will expose the fundamental entities from which the entire physical world is built.

While this exhilarating reductionist quest continues, progress occurs on the opposite front, at the interface of physics and biology, where the goal is to understand not what things are made of but how they are put together and function as integrated wholes. Here, the key concepts are complexity rather than simplicity, and organization rather than hardware. What is sought is a general 'Theory of Organization', or TOO.

Both TOEs and TOOs will undoubtedly lead to major revisions of known physics. TOEs have thrown up strange new ideas like the existence of extra space dimensions and the possibility that the world might be built out of strings – ideas which demand new areas of mathematics for their implementation. Likewise TOOs promise to uncover entirely new principles that will challenge the scope of existing physics.

The central issue facing the seekers of TOOs is whether the surprising – one might even say unreasonable – propensity for matter and energy to self-organize 'against the odds' can be explained using the known laws of physics, or whether completely new fundamental principles are required.

In practice, attempts to explain complexity and self-organization using the basic laws of physics have met with little success. In spite of the fact that the trend towards ever-greater organizational complexity is a conspicuous feature of the universe, the appearance of new levels of organization is frequently regarded as a puzzle, because it seems to go 'the wrong way' from a thermodynamic point of view. Novel forms of self-organization are therefore generally unexpected and prove to be something of a curiosity.

When presented with organized systems, scientists are sometimes able to model them in an *ad hoc* way after the fact. There is always considerable difficulty, however, in understanding how they came to exist in the first place, or in predicting entirely new forms of complex organization. This is especially true in biology. The origin of life, the evolution of increasing biological complexity, and the development of the embryo from a single egg cell, all seem miraculous at first sight, and all remain largely unexplained.

Nature's mysterious organizing power

Because of the evident problems in understanding complexity and self-organization in the universe there is no agreement on the source of nature's organizing potency. One can distinguish three different positions.

Complete reductionism

Some scientists assert that there *are* no emergent phenomena, that ultimately all physical processes can be reduced to the behaviour of elementary particles (or fields) in interaction. We are, they concede, at liberty to identify higher levels of description, but this is purely a convenience based on entirely subjective criteria. It is obviously far simpler to study a dog as a dog rather than a collection of cells, or atoms, interacting in a complicated way. But this practice must not fool us into thinking that 'dog' has any fundamental significance that is not already contained in the atoms that constitute the animal.

An extreme reductionist believes that all levels of complexity can in principle be explained by the underlying laws of mechanics that govern the behaviour of the fundamental fields and particles of physics. In principle, then, even the existence of dogs could be accounted for this way. The fact that we cannot in practice explain, say, the origin of life, is attributed solely to our current state of ignorance about the details of the complicated processes involved. But gaps in our knowledge must not, they caution, be filled by mysterious new forces, laws or principles.

My own position has been made clear in the foregoing chapters. Complete reductionism is nothing more than a vague promise founded on the outdated and now discredited concept of determinism. By ignoring the significance of higher levels in nature complete reductionism simply dodges many of the questions about the world that are most interesting to us. For example, it denies that the arrow of time has any reality. Defining a problem away does not explain it.

Uncaused creativity

Another point of view is to eschew reductionism in its most extreme form and admit that the existence of complex organized forms, processes and systems does not inevitably follow from the lower level laws. The existence of some, or all, higher levels is then simply accepted as a fact of nature. These new levels of organization (e.g. living matter) are not, according to this view, caused or determined in any way, either by the underlying levels, or anything else. They represent true novelty.

This was the position of the philosopher Henri Bergson. A teleologist, Bergson nevertheless rejected the idea of finalism as merely another form of determinism, albeit inverted in time:[1]

> The doctrine of teleology, in its extreme form, as we find it in Leibniz, for example, implies that things and beings merely realize a programme previously arranged. But there is nothing unforeseen; time is useless again. As in the mechanistic hypothesis, here again it is supposed *all is given*. Finalism thus understood is only inverted mechanism.

Bergson opts instead for the concept of a continuously creative universe, in which wholly new things come into existence in a way that is completely independent of what went before, and which is not constrained by a predetermined goal.

The concept of unrestrained creativity and novelty is also proposed by the modern philosopher Karl Popper:[2]

> Today some of us have learnt to use the word 'evolution' differently. For we think that evolution – the evolution of the universe, and especially the evolution of life on earth – has produced new things; *real novelty* . . . The story of evolution suggests that the universe has never ceased to be creative, or 'inventive'.

Some physicists have concurred with these ideas. For example, Kenneth Denbigh in his book *An Inventive Universe* writes:[3]

> Let us ask: Can genuinely new things come into existence during the course of time; things, that is to say, which are not entailed by the properties of other things which existed previously?

After outlining how this can indeed be the case, Denbigh addresses the question 'if the emergence of a new level of reality is always indeterminate, what is it 'due to', as we say?' He asserts that it has no cause at all:[4]

> The very fact that this kind of question seems to force itself on our attention shows the extent to which deterministic modes of thought have become deeply ingrained.

Denbigh prefers to think of the coming-into-being of new levels as an 'inventive process', that is, it brings into existence something which is both different and not necessitated: 'for the essence of true novelty is that *it did not have to be that way*'.

The difficulty I have in accepting this position is that it leaves the systematic nature of organization completely unexplained. If new organizational levels just pop into existence for no reason, why do we see such an orderly progression in the universe from featureless origin to rich diversity? How do we account for the regular progress of, say, biological evolution? Why should a collection of things which have no causes cooperate to produce a time-asymmetric sequence?

To say that this orderly unidirectional progression is uncaused, but just happens to be that way seems to me like saying that objects are not caused to fall by the force of gravity – they just happen to move that way. Such a point of view can never be called scientific, for it is the purpose of science to provide rational universal principles for the explanation of *all* natural regularities.

This brings me to the third alternative.

Organizing principles

If we accept that there exists a propensity in nature for matter and energy to undergo spontaneous transitions into new states of higher organizational complexity, and that the existence of these states is not fully explained or predicted by lower level laws and entities, nor do they 'just happen' to arise for no particular reason, then it is necessary to find some physical principles additional to the lower level laws to explain them.

I have been at pains to argue that the steady unfolding of organized complexity in the universe is a fundamental property of nature. I have reviewed some of the important attempts to model complex structures and processes in physics, chemistry, biology, astronomy and ecology. We have seen how spontaneous self-organization tends to occur in far-from-equilibrium open non-linear systems with a high degree of feedback. Such systems, far from being unusual, are actually the norm in nature. By contrast the closed linear systems studied in traditional mechanics, or the equilibrium systems of standard thermodynamics, are idealizations of a very special sort.

As more and more attention is devoted to the study of self-organization and complexity in nature, so it is becoming clear that there must be new general principles – organizing principles over and above the known laws of physics – which have yet to be discovered. We seem to be on the verge of discovering not only wholly new laws of nature, but ways of thinking about nature that depart radically from traditional science.

Software laws

What can be said about these new 'laws of complexity' and 'organizing principles' that seem to be the source of nature's power to create novelty? Talk of 'organizing principles' in nature is often regarded as shamefully mystical or vitalistic, and hence by definition anti-scientific. It seems to me, however, that this is an extraordinary prejudice. There is no compelling reason why the fundamental laws of nature have to refer only to the lowest level of entities, i.e. the fields and particles that we presume to constitute the elementary stuff from which the universe is built. There is no logical reason why new laws may not come into operation at each emergent level in nature's hierarchy of organization and complexity.

The correct position has been admirably summarized by Arthur Peacocke:[5]

> Higher level concepts and theories often refer to genuine aspects of reality at their own level of description and we have to eschew any assumptions that only the so-called fundamental particles of modern physics are 'really real'.

Let me dispel a possible misconception. It is not necessary to suppose that these higher level organizing principles carry out their marshalling of the system's constituents by deploying mysterious new forces specially for the purpose, which would indeed be tantamount to vitalism. Although it is entirely possible that physicists may discover the existence of new forces, one can still envisage that collective shepherding of particles takes place entirely through the operation of familiar inter-particle forces such as electromagnetism. In other words, the organizing principles I have in mind could be said to *harness* the existing interparticle forces, rather than supplement them, and in so doing alter the collective behaviour in a holistic fashion. Such organizing principles need therefore in no way contradict the underlying laws of physics as they apply to the constituent components of the complex system.

It is sometimes said that it is not possible to have organizing principles additional to the underlying (bottom level) laws of physics without contradicting those laws. Conventional physics, it is claimed, does not leave room for additional principles to act at the collective level. It is certainly true that laws at different levels can only co-exist if the system of interest is not over-determined. It is essential that the lower level laws are not in themselves so restrictive as to fix everything. To avoid this it is necessary to abandon strict determinism. It should be clear, however, from what has gone before, that strict determinism no longer has any place in science.

A word should be said about the use of the word 'law'. Generally speaking a law is a statement about any sort of regularity found in nature. The physicist, however, sets great store by those laws that apply with mathematical precision. A really hard-nosed reductionist would simply deny the existence of any other sort of law, claiming that *all* regularities in nature ultimately derive from a fundamental set of such mathematical laws. These days, that means some sort of fundamental Lagrangian from which a set of differential equations may be obtained.

With this restrictive usage, a law can only be tested by applying it to a collection of identical systems. As we come to consider systems of greater and greater complexity, the concept of a class of identical systems becomes progressively less relevant because an important quality of a very complex system is its uniqueness. It is doubtful, then, whether any mathematically exact statements can be made about classes of very complex systems. There can be no theoretical biology, for example, founded upon exact mathematical statements in the same way as in theoretical physics.

On the other hand, when dealing with complexity, it is the qualitative rather than the quantitative features which are of interest. The general trend towards increasing richness and diversity of form found in evolutionary biology is surely a basic fact of nature, yet it can only be crudely quantified, if at all. There is not the remotest evidence that this trend can be derived from the fundamental laws of mechanics, so it deserves to be called a fundamental law in its own right. In which case, it means accepting a somewhat broader definition of law than that hitherto entertained in physics.

The living world is full of regularities of this general, somewhat imprecise sort. For example, as far as I know all members of the animal kingdom have an even number of legs. It would be foolish to say that tripodal animals are *impossible*, but their existence is at least strongly suppressed. I am not suggesting that this 'law of the limbs' is in any sense basic. It may be the case, though, that such facts follow from a fundamental law regarding the nature of organized complexity in biology.

Many writers have used the example of the computer to illustrate the fact that a set of events might have two complementary and consistent descriptions at different conceptual levels – the hardware and the software. Every computer user knows that there can be 'software laws' that co-exist perfectly well with the 'hardware laws' that control the computer's circuitry. Nobody would claim that the laws of electromagnetism can be used to derive the tax laws just because the latter are stored in the Inland Revenue's computer!

We are therefore led to entertain the possibility that there exist 'software laws' in nature, laws which govern the behaviour of organization, information and complexity. These laws are fundamental, in the sense that they cannot logically be derived from the underlying 'hardware laws' that are the traditional subject matter of fundamental physics, but they are also compatible with those underlying laws in the same way that the tax laws can be compatible with the laws of electromagnetism. The software laws apply to *emergent* phenomena, inducing their appearance and controlling their form and behaviour.

Such ideas are by no means new. Many scientists and philosophers have argued that the laws of physics as presently conceived are inadequate to deal with complex organized systems – especially living systems. Moreover, these misgivings are not restricted to vitalists such as Dreisch. Even anti-vitalists point out that the reduction of all phenomena to the known laws of physics cannot wholly succeed because it fails to take into account the existence of different *conceptual levels* involved with complex phenomena.

In talking about biological organisms, for example, one wishes to make use of concepts such as teleonomy and natural selection, which are quite simply meaningless at the level of the physics of individual atoms. Biological systems possess a heirarchy of organization. At each successive level in the heirarchy new concepts, new qualities and new relationships arise, which demand new forms of explanation.

This point has been well expressed by the Cambridge zoologist W. H. Thorpe:[6]

> The behaviour of large and complex aggregates of elementary particles, so it turns out, is not to be understood as a simple extrapolation of the properties of a few particles. Rather, at each level of complexity entirely new properties appear, and the understanding of these new pieces of behaviour requires research which is as fundamental as, or perhaps even more fundamental than, anything undertaken by the elementary particle physicists.

This sentiment is not merely a jibe at the physics community. It is echoed by physicist P. W. Anderson, who writes:[7]

> I believe that at each level of organization, or of scale, types of behaviour open up which are entirely new, and basically unpredictable from a concentration on the more and more detailed analysis of the entities which make up the objects of these higher level studies.

The biologist Bernhard Rensch adopts a similar position:[8]

> We must take into consideration that chemical and biological processes, leading to more complicated stages of integration, also show the effects of *systemic* relations which often produce totally new characteristics. For example, when carbon, hydrogen and oxygen become combined, innumerable compounds can originate with new characteristics like alcohols, sugars, fatty acids,

and so on. Most of their characteristics cannot be deduced directly from the characteristics of the three basic types of atoms, although they are doubtless casually determined . . . We must ask now whether there are biological processes which are determined not only by causal but also by other laws. In my opinion, we have to assume that this is the case.

As mentioned in Chapter 8, dialectical materialism proposes similar ideas. We saw earlier how Engels believed that the second law of thermodynamics would actually be circumvented. Oparin apparently drew upon communist philosophy in support of his views concerning the origin of life:[9]

> According to the dialectic materialist view, matter is in a constant state of motion and proceeds through a series of stages of development. In the course of this progress there arises ever newer, more complicated and more highly evolved forms having new properties which were not previously present.

Biologist and Nobel prizewinner Sir Peter Medawar[10] has drawn an interesting parallel between the emerging conceptual levels in physics and biology and the levels of structure and elaboration in mathematics. In constructing the concepts of geometry, for example, the most primitive starting point is that of a topological space. This is a collection of points endowed with only very basic properties such as connectedness and dimensionality. Upon this meagre foundation one may first construct projective properties, enabling the concept of straight lines to be developed. Then one can build up so-called affine properties, which endow the space with a primitive form of directionality, and finally a metric may be imposed that gives full meaning to the concepts of distance and angle. The whole apparatus of geometrical theorems may then be constructed.

It would be absurd, Medawar points out, to talk of 'reducing' metrical geometry to topology. Metrical geometry represents a higher level *enrichment* of topology, which both contains the topological properties of the space and elaborates upon them. He sees this relationship between mathematical levels in a hierarchy of enrichment as paralleled in biology. Starting with atoms, building up through molecules, cells and organisms to conscious individuals and society, each level contains and enriches the one below, but can never be reduced to it.

Biotonic laws

What form, then, does this enrichment take in the case of biological systems? As I have already emphasized, one distinctive quality of all very complex systems, animate and inanimate, is their uniqueness. No two living creatures are the same, no two convection cell patterns are the same. We therefore have to contend with the problem of *individuality*. This point has been emphasized by a number of writers. Giuseppe Montalenti, for example, remarks that:[11]

> As soon as individuality appears, unique phenomena originate and the laws of physics become inadequate to explain all the phenomena. Certainly they are still valid for a certain number of biological facts, and they are extremely useful in explaining a certain number of basic phenomena; but they cannot explain everything. Something escapes them, and new principles have to be established which are unknown in the inorganic world: first of all natural selection, which gives rise to organic evolution and hence to life.

Montalenti is, however, anxious to dispel the impression that he is suggesting new and mysterious vital forces:

> This does not imply by any means either the introduction of vital forces or other metaphysical entities, nor does it mean that we should abandon the scientific method. The explanations we are looking for are always in the form of a cause–effect relationship, thus strictly adhering to scientific criteria; but the 'causes' and 'forces' implied are not only those known to physicists. Again, the example of natural selection, which is unknown in the physical world, is the most fitting. Others may be easily found in physiological, embryological and social phenomena.

A clear distinction between on the one hand espousing vitalism and on the other denying the reducibility of nature to the bottom level laws of physics is also made by Peacocke:[12]

> It *is* possible for higher level concepts and theories . . . to be non-reducible to lower level concepts and theories, that is, they can be autonomous. At the same time one has to recognize the applicability of the lower level concepts and theories (for example, those of physics and chemistry) to the component units of more complex entities and their validity when referred to that lower level. That is, with reference to biology, it is possible to be anti-reductionist without being a vitalist.

Similar views have been developed by the physicist Walter Elsasser, who lays great stress on the fact that living organisms are unique individuals and so do not form a homogeneous class suitable for study via the normal statistical methods of physics. This, he maintains, opens the way to the possibility of new laws, which he calls 'biotonic', that act at the holistic level of the organism, yet without in any way conflicting with the underlying laws of physics that govern the affairs of the particles of which the organism is composed:[13]

> We shall say at once that we accept basic physics as completely valid in its application to the dynamics of organisms ... Still, we must be clearly prepared to find that general laws of biology which are not deducible from physics will have *a logical structure quite different from what we are accustomed to* in physical science.

> To be specific, then, we assume that there exist regularities in the realm of organisms whose existence cannot be logico-mathematically derived from the laws of physics, nor can a logico-mathematical contradiction be construed between these regularities and the laws of physics.

The quantum physicist Eugene Wigner (also a Nobel prizewinner) likewise admits his 'firm conviction of the existence of biotonic laws'. He asks:[14] 'Does the human body deviate from the laws of physics, as gleaned from the study of inanimate matter?' and goes on to give two reasons involving the nature of consciousness why he believes the answer to be yes. One of these concerns the role of the observer in quantum mechanics, a topic to be discussed in Chapter 12. The other is the simple fact that, in physics, action tends to provoke reaction. This suggests to Wigner that, because matter can act on mind (e.g. in producing sensations) so too should mind be able to react on matter. He cautions that biotonic laws could easily be missed using the traditional methods of scientific investigation:[15]

> The possibility that we overlook the influence of biotonic phenomena, as one immersed in the study of the laws of macroscopic mechanics could have overlooked the influence of light on his macroscopic bodies, is real.

Another distinguishing characteristic of life is, of course, the teleological quality of organisms. It is hard to see how these can ever be reduced to the fundamental laws of mechanics. This view is also expressed in a recent review of teleology in modern science by astrophysicists John Barrow and

Frank Tipler, who write:[16] 'We do not think teleological laws either in biology or physics can be fully reduced to non-teleological laws.'

Downward causation

A further distinctive and important quality of all living organisms, emphasized by physicist Howard Pattee, is the *hierarchical* organization and control of living organisms. As smaller units integrate and aggregate into larger units, so they give rise to new rules which in turn constrain and regulate the component subsystems to comply with the collective behaviour of the system as a whole. This feature of higher levels in a hierarchy of organization acting to constrain lower levels of the same system is not restricted to biology. Pattee points out that a computer obeys all the laws of mechanics and electricity, yet no physicist would consider this statement a satisfactory answer to the question: 'what is the secret of a computing machine?' Pattee writes:[17]

> If there is any problem in the organization of a computer, it is the unlikely constraints which, so to speak, harness these laws and perform highly specific and directive functions.

The action of higher levels on lower levels has been called 'downward causation' by psychologist Donald Campbell, who remarks that:[18] 'all processes at the lower levels of a hierarchy are restrained by and act in conformity to the laws of the higher levels'.

There are a great many examples of downward causation in other branches of science. Karl Popper has pointed out that many of the devices of modern optics – lasers, diffraction gratings, holograms – are large-scale complex structures which constrain the motions of individual photons to conform with a coherent pattern of activity. In engineering, simple feedback systems engage in downward causation, as when a steam governor controls the flow of water molecules. Even the use of tools such as wedges can be viewed in terms of a macroscopic structure as a whole guiding the motion of its atomic constituents so as to produce, in concert, a particular result.

Similar ideas have been discussed by Norbert Weiner and E. M. Dewan in connection with control systems engineering.[19] In this subject a useful concept is that of *entrainment*, which occurs when an oscillator of some sort 'locks on to' a signal and responds in synchronism. A simple example of entrainment in action concerns tuning a television set. A detuned set will cause the picture to 'rotate', but if the frequency is adjusted the picture 'locks on' and stabilizes.

It was discovered 300 years ago by the physicist Huygens, the inventor of the pendulum clock, that if two clocks are mounted on a common support they will tick in unison. Such 'sympathetic vibrations' are now very familiar in the physics of coupled oscillators, which settle into 'normal modes' of vibration wherein all the oscillators execute collective synchronous patterns of motion. Cooperative vibration modes occur, for example, in crystal lattices, where each atom acts as a tiny oscillator. The propagation of light waves through crystals depends crucially on this organized collective motion.

Entrainment also occurs in electrical oscillators. If a power grid is supplied by a single generator, the frequency of supply is likely to drift due to variations in the generator output. If many generators are coupled into the grid, however, mutual entrainment stabilizes the oscillation frequency by pulling any drifting generator back into line. This tendency for coupled oscillators to 'beat as one' provides a beautiful example of how the behaviour of the system as a whole constrains and guides its individual components to comply with a coherent collective pattern of activity. The ability of such systems to settle down into collective modes of behaviour is one of the best illustrations of self-organization and is, of course, also the basis of the laser's remarkable self-organizing capability.

Laws of complexity and self-organization

The foregoing discussions show that one can distinguish between rather different sorts of organizing principles. It is convenient to refer to these as *weak, logical* and *strong*. Weak organizing principles are statements about the general way in which systems tend to self-organize. These include information about external constraints, boundary conditions, initial conditions, degree of non-linearity, degree of feedback, distance from equilibrium and so on. All of these facts are highly relevant in the various examples of self-organization so far discussed, yet they are *not* contained in the underlying laws themselves (unless one takes the extreme reductionist position). At present such statements are little more than a collection of *ad hoc* conditions and tendencies because our understanding of self-organizing phenomena is so rudimentary. It may not be too much to expect, however, that as this understanding improves some rather general and powerful principles will emerge.

Logical principles governing organization can be expected to come from the study of fractals, cellular automata, games theory, network theory, complexity theory, catastrophe theory and other computational models of complexity and information. These principles will be in the form of logical rules and theorems that are required on mathematical grounds. They will not refer to specific physical mechanisms for their proof. Consequently they will augment the laws of physics in helping us to describe organizational complexity.

A good example of a logical organizing principle is the universal appearance of Feigenbaum's numbers in the approach to chaos. These numbers arise for mathematical reasons, and are independent of the detailed physical mechanisms involved in producing chaos. Another would be the 'biological universals' discussed by Kauffman (see page 115) which attribute certain common emergent biological properties, not to shared descent and natural selection, but to the logical and mathematical relationships of certain automaton rules that govern wide classes of organic processes. Yet another are the hoped-for universal principles of order discussed by Wolfram in connection with cellular automata (see page 67).

Strong organizing principles are invoked by those who find existing physical laws inadequate to explain the high degree of organizational potency found in nature and see this as evidence that matter and energy are somehow being guided or encouraged into progressively higher organizational levels by additional creative influences. Such principles may be prompted by the feeling that nature is unusually efficient at conquering its own second law of thermodynamics and bringing about organized complexity. The origin of life and the origin of consciousness are often cited as examples that seem 'too good to be true' on the basis of chance and hint at some 'behind the scenes' creative activity.

There are two ways in which strong organizing principles can be introduced into physics. The first is to augment the existing laws with new principles. This is the approach of Elsasser, for example. The more radical approach is to modify the laws of physics. In the next chapter we shall examine some of these ideas.

I I

Organizing Principles

Cosmic principles

No scientist would claim that the existing formulation of the laws of physics is complete and final. It is therefore legitimate to consider that extensions or modifications of these laws may be found, that embody at a fundamental level the capacity for matter and energy to organize themselves. Many distinguished scientists have proposed such modifications, which have ranged from new cosmological laws at one extreme, to reformulations of the laws of elementary particles at the other.

Perhaps the best-known example of an additional organizing principle in nature is the so-called *cosmological principle*, which asserts that matter and radiation are distributed uniformly in space on a large scale. As we saw in Chapter 9 there is good evidence that this is the case. Not only did the matter and energy which erupted from the big bang contrive to arrange itself incredibly uniformly, it also orchestrated its motion so as to expand at exactly the same rate everywhere and in all directions. This uncanny conspiracy to create global order has baffled cosmologists for a long while.

The cosmological principle is really only a statement of the fact of uniformity. It gives no clue as to *how* the universe achieved its orderly state. Some cosmologists have been content to explain the uniformity by appealing to special initial conditions (i.e. invoking a weak organizing principle), but this is hardly satisfactory. It merely places responsibility for the uniformity with a metaphysical creation event beyond the scope of science.

An alternative approach has been a search for physical processes in the very early stages of the universe that could have had the effect of smoothing out an initially chaotic state. This idea is currently very popular, especially in the form of the inflationary universe scenario briefly described in Chapter 9. Nevertheless, whilst inflation does have a dramatic smoothing effect, it still requires certain special conditions to operate. Thus one continues to fall back on the need for either God-given initial conditions, or a cosmological organizing principle in addition to the usual laws of physics.

What can be said about such a principle? First it would have to be essentially acausal, or non-local, in nature. That is, the orchestration in the behaviour of regions of the universe that are spatially well separated requires *synchronized* global matching. There can be no time, therefore, for physical influences to propagate between these regions by any causative mechanism. (The theory of relativity, remember, forbids faster-than-light propagation of physical effects.)

Second, the principle can only refer to *large scale* organization, because on a scale less than galactic size uniformity breaks down. Here one recalls that the origin of the relatively small scale *irregularities* that gave rise to galaxies and clusters of galaxies is equally as mysterious as the large scale regularity of the cosmos. It is conceivable that the same cosmological organizing principle might account for both regularity and irregularity in the universe.

A suggestion for a possible new organizing principle has come from Roger Penrose, who believes that the initial smoothness of the universe ought to emerge from a time-asymmetric fundamental law. It is worth recalling at this stage that the second law of thermodynamics is founded upon the time-reversibility of the underlying system dynamics. If this is broken, the way is open to entropy reduction and spontaneous ordering. We have already seen how this happens in cellular automata, which undergo self-organization and entropy reduction. Penrose suggests something similar for cosmology.

It might be objected that physics has always been constructed around time-symmetric fundamental laws, but this is not quite true. Penrose points to the existence of certain exotic particle physics processes that display a weak violation of time reversal symmetry, indicating that at some deep level the laws of physics are not exactly reversible.

The details of Penrose's idea have been touched upon already at the end of Chapter 9. He prefers to characterize the smoothness of the early universe in terms of something called the Weyl curvature, which is a measure of the distortion of the cosmic geometry away from homogeneity and isotropy; crudely speaking, the Weyl tensor quantifies the clumpiness of the universe. The new principle would then have the consequence that the Weyl curvature is zero for the initial state of the universe. Such a state could be described as having very low gravitational entropy. As stressed in Chapter 9, more and more clumpiness (Weyl curvature) develops as the universe evolves, perhaps leading to black holes with their associated high gravitational entropy.

A more powerful cosmic organizing principle was the so-called *perfect cosmological principle*, this being the foundation of the famous steady-state theory of the universe due to Herman Bondi, Thomas Gold and Fred Hoyle. The perfect cosmological principle states that the universe, on a large scale, looks the same not only at all locations but also at all epochs. Put simply, the universe remains more or less unchanging in time, in spite of its expansion.

To achieve the perfect cosmological principle, its inventors proposed that the universe is forever replenishing itself by the continual creation of matter as it expands. The heat death of the universe is thereby avoided, because the new matter provides an inexhaustible supply of negative entropy. The ongoing injection of negative entropy into the universe was explained by Hoyle in terms of a so-called creation field, which had its own dynamics, and served to bring about the creation of new particles of matter at a rate which was automatically adjusted by the cosmological expansion. The universe thus became a huge self-regulating, self-sustaining mechanism, with a capacity to self-organize *ad infinitum*. The unidirectional character of increasing cosmic organization with time derives in this theory from the expansion of the universe, which drives the creation field and thereby provides an external arrow of time. Whatever its philosophical appeal the perfect cosmological principle has been undermined by astronomical observation.

Another well-known cosmic organizing principle is called Mach's principle after the Austrian philosopher Ernst Mach, though its origins go back to Newton. It is founded on the observation that although just about every object in space rotates, the universe as a whole shows no observable sign of rotation. Mach believed he had found the reason for this. He argued that the material contents of the universe as a whole serve to determine the local 'compass of inertia' against which mechanical accelerations are gauged, so by definition the universe cannot possess global rotation.

It is usually supposed that this coupling of local physics to the global distribution of cosmic matter is gravitational in nature. However, the dynamical laws of our best theory of gravitation – Einstein's general theory of relativity – does not embody Mach's principle (which really takes the form of a choice of boundary conditions). Such a principle cannot, therefore, be reduced to the gravitational field equations. It is an irreducibly non-local principle, additional to the laws of physics, that organizes matter in a cooperative way on a global scale.

Mach's principle is not the only example of this sort. Penrose has proposed a 'cosmic censorship hypothesis', which states that spacetime singularities that form by gravitational collapse must occur inside black holes; they can never be 'naked'. Another example is the 'no time travel' rule: gravitational fields can never allow an object to visit its own past.

Attempts to derive these restrictions from general relativity have not met with success, yet both are very reasonable conjectures. Indeed, if either naked singularities or travel into the past were permitted in the universe, it is hard to see how one could make any sense of physics. In both cases the restriction is of a global rather than local nature (black holes can only be properly defined in global terms). It therefore seems likely that some additional global organizing principle is required.

Microscopic organizing principles

A proposal to modify instead the *microscopic* laws of physics has been made by Ilya Prigogine. He points out that the inherent time symmetry of the laws of mechanics imply that they will never, as formulated, account for the time-asymmetric growth of complexity:[1]

> If the world were built like the image designed for reversible, eternal systems by Galileo Galilei and Isaac Newton, there would be no place for irreversible phenomena such as chemical reactions or biological processes.

His suggestion is to modify the laws of dynamics by introducing an intrinsic indeterminism reminiscent of quantum mechanics, but going beyond, in a way that is explicitly time-asymmetric. In that case

> the basic level of physics would be formed by nonequilibrium ensembles, which are less well determined than trajectories or [quantum] wave functions, and which evolve in the future in such a way as to increase this lack of determination.

Prigogine has developed an extensive body of mathematical theory in which he introduces this modification into the laws of dynamics, making them irreversible at the lowest, microscopic level. In this respect his proposal is similar to that of Penrose, discussed above. (For those interested in technicalities, Prigogine introduces non-Hermitian operators leading to non-unitary time evolution. The density matrix is acted upon by a superoperator that lifts the distinction between pure and mixed states, leading to the possibility of a rise in the system's microscopic entropy as it evolves.) He claims that the way is now open for understanding complexity in general, and explaining how order arises progressively out of chaos:[2]

> Most systems of interest to us, including all chemical systems and therefore all biological systems, are time-oriented on the macroscopic level. Far from being an 'illusion' [as the complete reductionist would claim], this expresses a broken time-symmetry on the microscopic level. Irreversibility is either true at *all* levels or none. It cannot fly as if by a miracle from one level to another. . .
>
> We come to one of our main conclusions: At all levels, be it the level of macroscopic physics, the level of fluctuations, or the microscopic level, nonequilibrium is the source of order. Nonequilibrium brings 'order out of chaos'.

A quite different suggestion for modifying the laws of physics in order to explain complex organization comes from the physicist David Bohm, a long-standing critic of the conventional interpretation of quantum mechanics. Bohm believes that quantum physics suggests an entirely new way of thinking about the subject of *order*, and is especially critical of the habit of equating randomness with disorder. He claims that randomness in quantum mechanics has actually been tested in only a few cases, and is by no means firmly established. But if quantum processes are not random, then the whole basis of neo-Darwinism is undermined:[3]

> We see, then, that even in physics, quantum processes may not take place in a completely random order, especially as far as short intervals of time are concerned. But after all, molecules such as DNA are in a continual process of rapid exchange of quanta of energy with their surroundings, so the possibility clearly exists that the current laws of quantum theory (based on the assumption of randomness of *all* quantum processes, whether rapid or slow) may be leading to seriously wrong inferences when applied without limit to the field of biology . . . It is evidently possible to go further and to assume that, under certain special conditions prevailing in the development of living matter, the order could undergo a further change, so that certain of these non-random features would be continued indefinitely. Thus there would arise *a new order of process*. The changes in this new order would themselves tend to be ordered in yet a higher order. This would lead not merely to the indefinite continuation of life, but to its indefinite evolution to an everdeveloping hierarchy of higher orders of structure and function.

Bohm is prepared to conjecture on specific ways in which non-randomness as it is manifested in biology might be tested:

One observation that could be relevant would be to trace a series of successive mutations to see if the order of changes is completely random. In the light of what has been said, it is possible that while a single change (or difference) may be essentially random relative to the previous state of a particular organism, there may be a tendency to establish a series of similar changes (or differences) that would constitute an *internally ordered* process of evolution.

He supposes that much of the time evolution is more or less random and so does not 'progress' but merely drifts stochastically, but these phases are punctuated by transitional periods of rapid, non-random change 'in which mutations tend to be fairly rapid and strongly directed in some order'. Such non-random behaviour would, claims Bohm, have very far-reaching consequences:

> for it would imply that when a given type of change had taken place there is an appreciable tendency in later generations for a series of similar changes to take place. Thus, evolution would tend to get 'committed' to certain general lines of development.

In the next chapter we shall see that many of the founding fathers of quantum mechanics believed that their new theory would cast important light on the mystery of living organisms, and many of them speculated about whether the theory would need to be modified when applied to biological phenomena. The belief that a modification of some sort is inevitable when applying the theory to the act of observation is shared by many physicists today.

A new concept of causation

A yet more radical reappraisal of the current formulation of physical laws has been proposed by theoretical biologist Robert Rosen. Rosen believes that the very concept of a physical law is unnecessarily restrictive, and in fact inadequate to deal with complex systems such as biological organisms:[4]

> The basis on which theoretical physics has developed for the past three centuries is, *in several crucial respects*, too narrow, and that far from being universal, the conceptual foundation of what we presently call theoretical physics is still very special; indeed, far too much so to accommodate organic phenomena (and much else besides).

Rosen points out that there is a traditional assumption among physicists that complex systems are merely special cases, i.e. complicated versions of simple systems. Yet as we have seen, there is increasing evidence that things are actually the other way around – that complexity is the norm and simplicity is a special case. We have seen how almost all dynamical systems, for example, belong to the unpredictable class called chaotic. The simple dynamical systems discussed in most physics textbooks, which have formed the principal topics in mechanics for 300 years, actually belong to an incredibly restricted class. Likewise in thermodynamics, the near-to-equilibrium closed systems presented in the textbooks are highly special idealizations. Much more common are far-from-equilibrium open systems.

It is no surprise, of course, that science has developed with this emphasis. Scientists naturally choose to work on problems with which they are likely to make some progress, and the above textbook examples are the ones that are most easily tackled. Complex systems are enormously harder to understand and are difficult to attack systematically. The spectacular progress made with simple systems has thus tended to obscure the fact that they are indeed very special cases.

This curious inversion of the traditional point of view leads Rosen to foresee that 'far from contemporary physics swallowing biology as the reductionists believe, biology forces physics to transform itself, perhaps ultimately out of all recognition'. He believes that physics must be considerably enlarged if it is to cope adequately with complex states of matter and energy.

Rosen gives as an example of the overly-restrictive conceptual basis of physics the assumption that all dynamical systems can be described by assigning them states which then evolve in accordance with dynamical laws. As explained in Chapter 2, this absolutely fundamental assumption lies at the heart of Newtonian dynamics, and was carried through to relativistic and quantum mechanics as well as field theory and thermodynamics. It is a formulation that embodies the very concept of causality as it has been understood for the last 300 years, and is closely tied to the conventional ideas of determinism and reversibility.

This key assumption, however, implies an extremely special sort of mathematical description. (Technically this has to do with the existence of exact differentials which ultimately derives from the existence of a Lagrangian.) Generally, if one has a set of quantities describing the rates of change of various features of a complex system, it will *not* be possible to combine these quantities in such a way as to recover the above-mentioned very special description. Rosen argues that the theory of dynamical systems should be enlarged to accommodate those cases where the special description fails. Such cases will actually, claims Rosen, represent the vast majority of systems found in nature. The restricted set of dynamical systems at present studied by physicists will turn out to belong to a very special class.

The changes which Rosen proposes, and which he has developed in quite some mathematical detail, amount to much more than technical tinkering. They demand a completely new vocabulary. Crucially, for example, the quantities that change will generally be *informational* in nature, so that Rosen explicitly introduces the idea that I have called software laws. He distinguishes *simple systems* of the type traditionally studied in physics (where states and dynamical laws in the form of differential equations constitute a highly idealized scheme) from *complex systems* 'describable by a web of informational interactions'. Of the former, Rosen says 'one can even question whether there are any simple systems at all; if there are not then our traditional universals evaporate entirely'.

A radical reformulation along such lines restores the old Aristotelian classes of causation, even leaving room for the notion of final causes:[5]

> Complex systems can allow a meaningful, scientifically sound category of final causation, something which is absolutely forbidden within the class of simple systems. In particular, complex systems may contain subsystems which act as predictive models of themselves and/or their environments, whose predictions regarding future behaviours can be utilized for modulation of present change of state. Systems of this type act in a truly anticipatory fashion, and possess novel properties.

Whilst fundamental modifications of the laws of physics of the sort proposed by Prigogine and Rosen must still be regarded as highly speculative, they show how the existence of complexity in nature is seen by some scientists to challenge the very basis on which the laws of science have been formulated.

Wilder ideas

One of the founders of quantum mechanics was Wolfgang Pauli, of Pauli exclusion principle fame. Pauli enjoyed an interesting association with the psychoanalyst Carl Jung, and helped Jung to develop a provocative concept that flies in the face of traditional ideas of causation.

It was Jung's contention that scientific thinking has been unreasonably dominated by notions of causality for the explanation of physical events.

He was impressed by the fact that quantum mechanics undermines strict causality, reducing it to a statistical principle, because in quantum physics events are connected only probabilistically. Jung therefore saw the possibility that there may exist alongside causality another physical principle connecting in a statistical way events that would otherwise be regarded as independent:[6]

> Events in general are related to one another on the one hand as causal chains, and on the other hand by a kind of *meaningful cross-connection.*

He called this additional principle *synchronicity.*

To establish whether or not synchronicity exists, Jung was led to examine the nature of *chance* events, to discover whether 'a chance event seems causally unconnected with the coinciding fact'.[7] He assembled a great deal of anecdotal evidence for exceedingly improbable coincidences, many taken from his own medical casework. The typical sort of thing is familiar to us all. You run into an old friend the very day you were talking about him. The number on your bus ticket turns out to be exactly the telephone number you just dialled. Jung considered some of these stories to be utterly beyond the bounds of coincidence as to constitute evidence for an acausal connecting principle at work:[8]

> All natural phenomena of this kind are unique and exceedingly curious combinations of chance, held together by the common meaning of their parts to form an unmistakable whole. Although meaningful coincidences are infinitely varied in their phenomenology, as acausal events they nevertheless form an element that is part of the scientific picture of the world. Casuality is the way we explain the link between two successive events. Synchronicity designates the parallelism of time and meaning between psychic and psychophysical events, which scientific knowledge has so far been unable to reduce to a common principle.

In spite of the popularization of Jung's ideas by Arthur Koestler in his book *The Roots of Coincidence*,[9] synchronicity has not been taken seriously by scientists. Probably this is because much of the evidence which Jung presented drew upon discredited subjects like astrology and extra-sensory perception. Most scientists prefer to regard stories of remarkable coincidences as a selection effect: we remember the occasional unexpected conjunction of events, but forget the myriad of unremarkable events that happen all the time. For every dream that comes true there are millions that do not. From time to time the odd dream must come true, and that will be the one which is remembered.

It is interesting, nevertheless, to consider from the point of view of physics what would be involved in a synchronicity principle. This is best discussed with reference to a spacetime diagram. In Figure 30 time is drawn as a vertical line and a single dimension of space as a horizontal line. A point on the diagram is called an *event*, because it is assigned both a place and a moment. A horizontal section through the diagram represents all space at one instant of time, and it is usual to think of time as flowing up the diagram, so that future is towards the top of the diagram and past is towards the bottom.

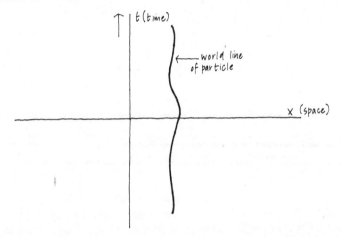

Figure 30. Space-time diagram. Points on the diagram represent events; the wiggly line represents the career of a particle through space and time.

The fact that the natural world is not merely a chaotic jumble of independent events, but is ordered in accordance with the laws of nature, imposes some order on the spacetime diagram. For example, the fact that an object such as an atom continues to exist as an identifiable entity through time means that it traces out a continuous path, or *world line* in spacetime. If the object moves about in space then the world line will be wiggly.

Figure 31 shows a number of world lines. In general the shapes of these lines will not be independent because there will be forces of interaction between the particles. The disturbance of one particle will have a *causative influence* on the others, and this will show up as *correlations* between events lying on neighbouring world lines. The rules governing cause and effect in spacetime are subject to the restrictions of the theory of relativity, which forbids any physical influence from propagating faster than the speed of light. The world line of a light pulse is an oblique straight line, which it is

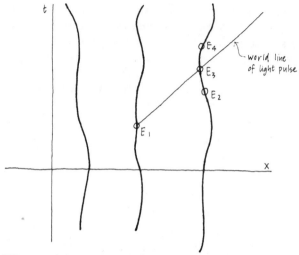

Figure 31. This space-time diagram shows the world lines of three material particles, as well as that of a pulse of light (oblique line). Such light paths determine which events can causally interact with other events. Thus E_1 can affect E_3 and E_4, but not E_2.

conventional to draw at $45°$. Thus, pairs of events such as E_1, E_2 cannot be causally connected because they lie in spacetime outside the region delimited by the light line through E_1. Such pairs of events are said to be *spacelike* separated. On the other hand E_1 can have a causative influence on E_3 or E_4. These events are *not* spacelike separated from E_1.

Although cause and effect cannot operate between spacelike separated events, that does not mean that events such as E_1 and E_2 must be completely unrelated to each other. It may be that both events are triggered by a common causative event that lies between them in space. This would occur, for example, if two light pulses were sent in opposite directions and caused the simultaneous detonation of two widely separated explosive charges. However, such was not what Jung had in mind with synchronicity.

In the next chapter we shall see how quantum mechanics permits the existence of correlations between simultaneous events separated in space which would be impossible in any classical picture of reality. These *non-local* quantum effects are indeed a form of synchronicity in the sense that they establish a connection – more precisely a correlation – between events for which any form of causal linkage is forbidden.

It is sufficient, but not necessary, for the elimination of causal connection that events are spacelike separated. It may happen that causal connection is permitted by relativity, but is otherwise improbable. Relativity does not forbid the discussion of a friend from causing his prompt appearance, but the causation seems unlikely.

More generally one can envisage *constellations* of events in spacetime, associated in some meaningful way, yet without causal association. These events may or may not be spacelike separated, but their conjunction or association might not be attributable to causal action. They would form patterns or groupings in spacetime representing a form of order that would not follow from the ordinary laws of physics. In fact, the sort of organizing principles discussed in the foregoing sections could be described in these terms, and regarded as a form of synchronicity. However, whereas acausal associations in, say, biosystems might be reasonable, it is quite another matter to extend the idea to events in the daily lives of people, which was Jung's chief interest.

Another set of 'meaningful coincidences' have recently attracted the attention of scientists. This time the coincidences do not refer to events but to the so-called constants of nature. These are numbers which crop up in the various laws of physics; examples include the mass of the electron, the electric charge of the proton and Newton's gravitational constant (which fixes the strength of the gravitational force). So far the values of these various constants are unexplained by any theory, so the question arises as to why they have the values that they do. Now the interesting thing is that the existence of many complex structures in the universe, and especially biological organisms, is remarkably sensitive to the values of the constants. It turns out that even slight changes from the observed values suffice to cause drastic changes in the structures. In the case of organisms, even minute tinkering with the constants of nature would rule out life altogether, at least of the terrestrial variety.

Nature thus seems to be possessed of some remarkable numerical coincidences. The constants of nature have, it appears, assumed precisely the values needed in order that complex self-organization can occur to the level of conscious individuals. Some scientists have been so struck by this contrivance, that they subscribe to something called the *strong anthropic principle*, which states that the laws of nature must be such as to admit the existence of consciousness in the universe at some stage. In other words, nature organizes itself in such a way as to make the universe self-aware. The strong anthropic principle can therefore be regarded as a sort of organizing meta-principle, because it arranges the *laws themselves* so as to permit complex organization to arise.

Another very speculative theory that goes outside the causal bounds of space and time has been proposed by biologist Rupert Sheldrake.[10] The central problem of this book – the origin of complex forms and structures – has been tackled by Sheldrake in a head-on fashion. In Chapter 7 it was mentioned that a fashionable idea in developmental biology is that of the *morphogenetic field*. These fields are invoked as an attempt to explain how an egg cell develops into a complicated three-dimensional structure. The nature and properties of morphogenetic fields remain somewhat uncertain, if indeed they exist at all.

Sheldrake proposes to take morphogenetic fields seriously, and interpret them as an entirely new type of physical effect. He believes that in some way the field stores the information about the final form of the embryo, and then guides its development as it grows. This seems, therefore, like a revival of old-fashioned teleology. Sheldrake injects a new element, however, in his hypothesis of *morphic resonance*. The idea is that once a new type of form has come into existence, it sets up its own morphogenetic field which then encourages the appearance of the same form elsewhere. Thus, once nature has 'learned' how to grow a particular organism it can guide, by 'resonance', the development of other organisms along the same pathway.

Morphogenetic fields are not, according to Sheldrake, restricted to living organisms. Crystals possess them too. That is why, he believes, there have been cases where substances which have previously never been seen in crystalline form have apparently been known to start crystallizing in different places at more or less the same time. Sheldrake's fields are also associated with memory. Once an animal has learnt to perform a new task, others find it easier to learn that task.

The fields which Sheldrake has in mind do not act in space and time in the usual causative fashion. Indeed, it has to be said that the nature of the fields is completely mysterious from the point of view of physics. Nevertheless, the theory at least has the virtue of falsifiability, and Sheldrake has proposed a number of experimental tests involving human learning. So far the results have proved inconclusive.

The rather bizarre ideas I have mentioned in this section do not form part of mainstream science and should not, perhaps, be taken very seriously. Nevertheless they illustrate the persistence of the impression among scientists and laymen alike that the universe has been organized in a way that is hard to explain mechanistically, and that in spite of the tremendous advances in fundamental science there is still a strong temptation to fall back on some higher principle.

12

The Quantum Factor

Quantum weirdness and common sense

It is often said that physicists invented the mechanistic-reductionist philosophy, taught it to the biologists, and then abandoned it themselves. It cannot be denied that modern physics has a strongly holistic, even teleological flavour, and that this is due in large part to the influence of the quantum theory.

When quantum mechanics was properly developed in the 1920s it turned science upside down. This was not only due to its astonishing success in explaining a wide range of physical phenomena. As with the theory of relativity which preceded it, quantum mechanics swept away many deeply entrenched assumptions about the nature of reality, and demanded a more abstract vision of the world.

Common sense and intuition were the first victims. Whereas the old physics generally employed everyday concepts of space, time and matter differing from familiar experience only in degree, the new physics was formulated in terms of abstract mathematical entities and algorithms. Attempts to cast what is 'going on' in the language of ordinary experience frequently appear mystical, absurd or even flatly paradoxical. We are saved from being assaulted by the madhouse of the quantum in our daily affairs only by virtue of the fact that quantum effects are generally limited to the submicroscopic realm of atoms, molecules and subatomic particles.

In classical mechanics the state of a system is easily visualized. It is given by specifying the positions and velocities (or momenta) of all the particles concerned. The system evolves as the particles move about under the influence of their mutual interactions and any externally applied forces. The physicist can predict this evolution, at least in principle, by the use of Newton's laws of motion to compute the paths in space of each particle.

Quantum mechanics replaces this concrete picture of the state of a mechanical system by an abstract mathematical object called the *wave function* or the *state vector*. This is not something that has any physical counterpart – it is not itself an observable thing. There is, however, a well-defined mathematical procedure for extracting information from the wave function about things that are observable (e.g. the position of a particle).

Where quantum mechanics differs fundamentally from classical mechanics is not so much in this 'one-step-removed' procedure than in the fact that the wave function only yields *probabilities* about observable quantities. For example, it is not generally possible, given the wave function, to predict *exactly* where a particle is located, or how it is moving. Instead, only the relative probabilities can be deduced that the particle is to be found in such-and-such a region of space with such-and-such a velocity.

Quantum mechanics is therefore a *statistical* theory. But unlike other statistical theories (e.g. the behaviour of stock markets, roulette wheels) its probabilistic nature is not merely a matter of our ignorance of details; it is inherent. It is not that quantum mechanics is inadequate to predict the precise positions, motions, etc. of particles; it is that a quantum particle simply *does not possess* a complete set of physical attributes with well-defined values. It is meaningless to even consider an electron, say, to have a precise location and motion at one and the same time.

The inherent vagueness implied by quantum physics leads directly to the famous uncertainty or indeterminacy principle of Werner Heisenberg, which states that pairs of quantities (e.g. the position and momentum of a particle) are incompatible, and cannot have precise values simultaneously. The physicist can choose to measure either quantity, and obtain a result to any desired degree of precision, but the more precisely one quantity is measured, the less precise the other quantity becomes.

In classical mechanics one must know *both* the positions and the momenta of all the particles at the same moment to predict the subsequent evolution of the system. In quantum mechanics this is forbidden. Consequently there is an intrinsic uncertainty or indeterminism in how the system will evolve. Armed even with the most complete information permitted about a quantum system it will generally be impossible to say what value any given quantity (e.g. the position of a particle) will have at a later moment. Only the betting odds can be given.

In spite of the indeterminism that is inherent in quantum physics, a quantum system can still be regarded as deterministic in a limited sense, because the *wave function* evolves deterministically. Knowing the state of the system at one time (in terms of the wave function), the state at a later time can be computed, and used to predict the relative probabilities of the values that various observables will possess on measurement. In this weaker form of determinism, the various probabilities evolve deterministically, but the observable quantities themselves do not.

The fact that in quantum physics one cannot know everything all of the time leads to some oddities. An electron, for instance, may sometimes behave like a wave and sometimes like a particle – the famous 'wave-particle duality'. Many of these weird effects arise because a quantum state can be a *superposition* of other states. Suppose, for example, there is a particular wave function, A, corresponding to an electron moving to the left, and another, B, corresponding to an electron moving to the right. Then it is possible to construct a quantum state described by a wave function which consists of A and B superimposed. The result is a state in which, in some sense, both left-moving and right-moving electrons co-exist or, more dramatically, in which two worlds, one containing a left-moving electron and the other a right-moving electron, are present together. Whether these two worlds are to be regarded as equally real, or merely alternative contenders for reality is a matter of debate. There is no disagreement, however, that superpositions of this sort often occur in quantum systems.

The ability of quantum objects to possess apparently incompatible or contradictory properties – such as being both a wave and a particle – prompted Niels Bohr, the Danish physicist who more than any other clarified the conceptual basis of the theory, to introduce his so-called principle of complementarity. Bohr recognized that it is not paradoxical for an electron to be *both* a wave and a particle because the wave-like and particle-like aspects are never displayed in a contradictory way in the same experiment. Bohr pointed out that one can construct an experiment to display the wave-like properties of a quantum object, and another to display its particle-like properties, but never both together. Wave and particle behaviour (and other incompatibilities, such as position and momentum) are not so much *contradictory* as *complementary* aspects of a single reality. Which face of the quantum object is presented to us depends on how we choose to interrogate it.

What happens to an atom when it's being watched?

Bohr's principle of complementarity demands a fundamental reappraisal of the nature of reality, in particular of the relationships between the part and the whole, and the observer and observed. Clearly, if an electron is to possess, say, either a well-defined position or a well-defined momentum dependent on which aspect of its reality one chooses to observe, then the properties of the electron are inseparable from those of the measuring apparatus – and by extension the experimenter – used to observe it. In other words, we can only make meaningful statements about the condition of an electron *within the context of a specified experimental arrangement*. No meaningful value can be attached, for example, to the position of a given electron at the moment we are measuring its momentum.

It follows that the state of the quantum microworld is only meaningfully defined with respect to the classical (non-quantum) macroworld. It is necessary that there *already exist* macroscopic concepts such as a measuring apparatus (at least in principle) before microscopic properties, such as the position of an electron, have any meaning.

There is a touch of paradox here. The macroworld of tables, chairs, physics laboratories and experimenters is *made up of* elements of the microworld: the measuring apparatus and experimeter are themselves composed of quantum particles. There is thus a sort of circularity involved: the macroworld needs the microworld to constitute it and the microworld needs the macroworld to define it.

The paradoxical nature of this circularity is thrown into sharp relief when the act of measurement is analysed. Although the microworld is inherently nebulous, and only probabilities rather than certainties can be predicted from the wave function, nevertheless when an actual measurement of some dynamical variable is made a concrete result is obtained. The act of measurement thus transforms probability into certainty by *projecting out* or *selecting* a specific result from among a range of possibilities. Now this projection brings about an abrupt alteration in the form of the wave function, often referred to as its 'collapse', which drastically affects its subsequent evolution.

The collapse of the wave function is the source of much puzzlement among physicists, for the following reason. So long as a quantum system is not observed, its wave function evolves deterministically. In fact, it obeys a differential equation known as the Schrödinger equation (or a generalization thereof). On the other hand, when the system is inspected by an external observer, the wave function suddenly jumps, in flagrant violation of Schrödinger's equation. The system is therefore capable of changing with time in two completely different ways: one when nobody is looking and one when it is being observed.

The rather mystical conclusion that observing a quantum system interferes with its behaviour led von Neumann to construct a mathematical model of a quantum measurement process.[1] He considered a model microscopic quantum system – let us suppose it is an electron – coupled to some measuring apparatus, which was itself treated as a quantum system. The whole system – electron plus measuring apparatus – then behaves as a large, integrated and closed quantum system that satisfies a super-Schrödinger equation. Mathematically, the fact that the system treated as a whole satisfies such an equation ensures that the wave function representing the entire system must behave deterministically, whatever happens to the part of the wave function representing the electron.

It was von Neumann's intention to find out how the coupled quantum dynamics of the whole system brings about the abrupt collapse of the electron's wave function. What he discovered was that the act of coupling the electron appropriately to the measuring device can indeed cause a collapse in that part of the wave function pertaining to our description of the electron, but that the wave function representing the *system as a whole* does not collapse.

The conclusion of this analysis is known as 'the measurement problem'. It is problematic for the following reason. If a quantum system is in a superposition of states, a definite reality can only be observed if the wave function collapses on to one of the possible observable states. If, having included the observer himself in the description of the quantum system, no collapse occurs, the theory seems to be predicting that there is no single reality.

The problem is graphically illustrated by the famous Schrödinger cat paradox. Schrödinger envisaged a cat incarcerated in a box with a flask of cyanide gas. The box also contains a radioactive source and a geiger counter that can trigger a hammer to smash the flask if a nucleus decays. It is then possible to imagine the quantum state of a nucleus to be such that after, say, one minute, it is in a superposition corresponding to a probability of one-half that decay has occurred and one-half that it has not. If the entire box contents, including the cat, are treated as a single quantum system, we are forced to conclude that the cat is also in a superposition of two states: dead and alive. In other words, the cat is apparently hung up in a hybrid state of unreality in which it is somehow both dead and alive!

Many attempts have been made to resolve the foregoing quantum measurement paradox. These range from the mystical to the bizarre. In the former category is Wigner's proposal, mentioned briefly in Chapter

10, that the mind of the experimenter (or the cat?) collapses the wave function:[2] 'It is the entering of an impression into our consciousness which alters the wave function . . . consciousness enters the theory unavoidably and unalterably.' In the bizarre category is the many universes interpretation, which supposes that all the quantum worlds in a superposition are equally real. The act of measurement causes the entire universe to split into all quantum possibilities (e.g. live cat, dead cat). These parallel realities co-exist, each inhabited by a different copy of the conscious observer.

Beyond the quantum

Attempts to escape from the quantum measurement paradox fall into two categories. There are those, such as the many-universes theory just described, that accept the universal validity of quantum mechanics as their starting point. Then there are the more radical theories, which conjecture that quantum mechanics breaks down somewhere between the micro- and macroworlds. This may occur at a certain threshold of size or, more convincingly, at a certain threshold of complexity. It has already been mentioned in Chapter 10 how David Bohm has questioned the true randomness of quantum events when applied to biosystems.

Of those who have suggested that quantum mechanics fails when applied to complex systems, perhaps the best known is Eugene Wigner, one of the founders of quantum mechanics. Wigner bolsters his claim with a mathematical analysis of biological reproduction, in which he considers a closed system containing an organism together with some nutrient.[3] By treating the system using the laws of quantum mechanics, he concludes that it is virtually impossible for the system to evolve in time in such a way that at a later moment there are two organisms instead of one. In other words, asserts Wigner, biological reproduction is inconsistent with the laws of quantum mechanics. This inconsistency is most conspicuously manifested, he says, during the act of quantum measurement, where it is the entry of information about the quantum system into the consciousness of the observer that brings about the collapse of the wave function.

Many of Wigner's ideas were shared by von Neumann, who was also sceptical about the validity of quantum mechanics when extended to organic phenomena. On one occasion von Neumann was engaged in debate with a biologist who was trying to convince him of the neo-Darwinist theory of evolution. von Neumann led the biologist to the window of his study and said, cynically:[4] 'Can you see the beautiful white villa over there on the hill? It arose by pure chance.' Needless to say, the biologist was unimpressed.

Another scientist who questions the universal validity of quantum mechanics is Roger Penrose. His scepticism comes from considerations of black holes and cosmology, as discussed in Chapter 10. He writes:[5]

> There is something deeply unsatisfactory about the present conventional formulation of quantum mechanics, incorporating as it does two quite distinct modes of evolution: one completely deterministic, in accordance with the Schrödinger equation, the other a probabilistic collapse. And it is a great weakness of the conventional theory that one is not told when one form of evolution is supposed to give way to the other, beyond the fact that it must always take place sometime prior to an observation being made ... if I am right, then Schrödinger's equation will have to be modified in some way.

Penrose suggests a modification which introduces a radical new proposal – that *gravitation* is in some way involved in the collapse of the wave function. He thus ties in his misgivings about the validity of quantum mechanics in the macroscopic world with his attempt to formulate a time-asymmetric law to explain the gravitational smoothness of the early universe. (It is worth recalling here that the collapse of the wave function is a time-asymmetric process.)

The reader will be convinced, I am sure, that the quantum measurement problem remains unresolved. There is, however, at least one point of agreement: an act of measurement can only be considered to have taken place when some sort of record or trace is generated. This could be a track in a cloud chamber, the click of a geiger counter or the blackening of a photographic emulsion. The essential feature is that an *irreversible* change occurs in the measuring apparatus which conveys meaningful information to the experimenter. Bohr spoke of 'an irreversible amplification' of the microscopic disturbance that triggers the measuring device, putting the device into a concrete state that can be 'described in plain language' (e.g. the counter has clicked, the pointer is in position 3). The upshot is that the concept of measurement must always be rooted in the classical world of familiar experience.

Quantum measurement as an example of downward causation

I have emphasized that the wave function description of the state of a quantum system is not a description of where the particles are and how they are moving, but something more abstract from which some statistical information about these things can be obtained. The wave function represents not how the system *is*, but what we *know* about the system.

Once this fact is appreciated, the collapse of the wave function is no longer so mysterious, because when we make a measurement of a quantum system our knowledge of the system changes. The wave function then changes (collapses) to accommodate this. On the other hand, the evolution of the wave function determines the relative probabilities of the outcomes of future measurements, so the collapse does have an effect on the subsequent behaviour of the system. A quantum system evolves differently if a measurement is made than if it is left alone.

Now this is not, in itself, so very bizarre. Indeed, the same is true of a classical system; whenever we observe something we disturb it a bit. In quantum mechanics, however, this disturbance is a fundamental, irreducible and unknowable feature. In classical mechanics it is merely an incidental feature: the disturbance can be reduced to an arbitrarily small effect, or computed in detail and taken into account. Such is not possible for a quantum system.

The act of quantum measurement is a clear example of downward causation, because something which is meaningful at a higher level (such as a geiger counter) brings about a fundamental change in the behaviour of a lower level entity (an electron, say). In fact, quantum measurement *unavoidably* involves downward causation because if we try to treat the measuring apparatus on the *same* level as the electron – by considering it as just a collection of quantum particles described by a total wave function – then, as we have seen, no measurement takes place. The very meaning of measurement refers to our drawing a distinction between the microscopic level of elementary particles and the macroscopic level of complex pieces of apparatus in which irreversible changes take place and traces are recorded.

The downward causation involved here can also be viewed in terms of information. The wave function, which contains our knowledge of the quantum system, may be said to represent information; in computer jargon, software. Thus the wave associated with, say, an electron, is a wave of *software*. On the other hand the particle aspect of an electron is akin to hardware. Using this language one might say that the quantum wave-particle duality is a hardware–software duality of the sort familiar in computing. Just as a computer has two complementary descriptions of the same set of events, one in terms of the program (e.g. the machine is working out somebody's tax bill) and another in terms of the electric circuitry, so the electron has two complementary descriptions – wave and particle.

In its normal mode of operation a computer does not, however, provide an example of downward causation. We would not normally say that the act of multiplication *causes* certain circuits to fire. There is merely a parallelism in the hardware and software descriptions of the same set of events. In the quantum measurement case, what is apparently a closed quantum system (electron plus measuring apparatus plus experimenter) evolves in such a way that there is a change in the information or software, which in turn brings about a change in the hardware (the electron moves differently afterwards).

I have tried to extend the computer analogy to cover this. Consider a computer equipped with a mechanism such as a robot arm, capable of moving about in accordance with a program in the computer. Such devices are familiar on car assembly lines. Now ask what happens if the computer is programmed so that the arm begins carrying out modifications *to the computer's own circuitry* (see Figure 32). This is an example of software–hardware feedback. Just as changes in information downwardly cause changes in the behaviour of an electron during a quantum measurement, so changes in the program software downwardly cause modifications in the computer's hardware.

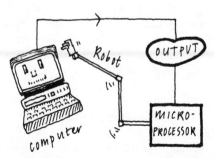

Figure 32. A computer programmed to rearrange its own circuitry provides as example of 'level mixing'. Software and hardware become logically entangled in a fashion suggestive of wave-particle duality in quantum physics.

The physicist John Wheeler gives an even more vivid interpretation of quantum measurement as downward causation.[6] 'How is it possible,' he asks, 'that mere information (that is: "software") should in some cases modify the real state of macroscopic things (hardware)?' To answer this question Wheeler first concurs with Bohr that a measurement requires some sort of irreversible amplification resulting in a record or trace, but in Wheeler's view this is not enough. He believes that a measurement can only be considered to have taken place when a *meaningful* record exists.

When is a record meaningful? Wheeler appeals to the rather abstract notion of a 'community of investigators' for whom a click on a geiger counter or deflection of a pointer *means* something. He traces a circuit of causation or action from elementary particles through molecules and macroscopic objects to conscious beings and communicators and meaningful statements, and urges us 'to abandon for the foundation of existence a physics hardware located "out there" and to put instead a meaning software'. In other words, meaning – or information, or software – is elevated to primary status, and particles of matter become secondary. Thus, proclaims Wheeler, 'Physics is the child of meaning even as meaning is the child of physics.'

However, now Wheeler asks about how the 'meaning circuit' is to be closed. This must involve some kind of *reaction* of meaning on the physical world of elementary particles – the 'return portion of the circuit'. Such downward causation is taken to be equally as fundamental, if as yet more obscure, than the 'upward' part of the circuit.

The details of the downward causation here remain enigmatic, save in one respect. The normal course of upward causation is forward in time (an atom decays, a particle emerges, a counter clicks, an experimenter reads the counter . . .). The return portion must therefore be 'backwards in time'. Wheeler illustrates this with a new experiment, called the delayed-choice experiment, which involves a type of retro-causation. The experiment has recently been conducted,[7] and accords entirely with Wheeler's expectations. It must be understood, however, that no actual communication with the past is involved.

Are the higher levels primary?

The crucial *irreversibility* involved in all interpretations of quantum measurement recalls Prigogine's philosophy that irreversible phenomena – the phenomena of becoming – are primary, while reversible processes –

the phenomena of being – are approximations or idealizations of a secondary nature. Quantum physics places observation (or at least measurement) at the centre of the stage of reality, while treating elementary particles as mere abstractions from these primary experiences (or events).

Physicists often talk informally about electrons, atoms and so on as though they enjoy a complete and independent existence with a full set of attributes. But this is a fiction. Quantum physics teaches us that electrons simply don't exist 'out there' in a well-defined sense, with places and motions, in the absence of observations. When a physicist uses the word 'electron' he is really referring to a mathematical algorithm which enables him to relate in a systematic way the results of certain very definite and precisely specified experiments. Because the relations are systematic it is easy to be seduced into believing that there really is a little thing 'out there', like a scaled-down version of a billiard ball, producing the results of the measurements. But this belief does not stand up to scrutiny.

Quantum physics leads to the conclusion that the bottom level entities in the universe – the elementary particles out of which matter is composed – are not really elementary at all. They are of a secondary, derivative nature. Rather than providing the concrete 'stuff' from which the world is made, these 'elementary' particles are actually essentially *abstract* constructions based upon the solid ground of irreversible 'observation events' or measurement records.

This seems to be Prigogine's own position:[8]

> The classical order was: particles first, the second law [of thermodynamics] later – being before becoming! It is possible that this is no longer so when we come to the level of elementary particles and that here we must *first* introduce the second law before being able to define the entities . . . after all, an elementary particle, contrary to its name, is not an object that is 'given'; we must construct it.

Prigogine recalls Eddington's division of laws into primary (such as Newton's laws of motion for individual particles) and secondary (such as the second law of thermodynamics). Eddington wondered if[9] 'in the reconstruction of the scheme of physics, which the quantum theory is now pressing on us, secondary laws become the basis and primary laws are discarded'. In other words, downward causation takes precedence over upward causation.

These considerations give quantum physics a strong holistic, almost

Aristotelian flavour. Here we find not only the whole being greater than the sum of its parts, but also the existence of the parts being defined by the whole in a gigantic hardware-software mixing of levels.

A physicist who has developed this theme in great detail, and drawn parallels with oriental philosophy, is David Bohm. Bohm sees quantum physics as the touchstone of a new conception of order and organization that extends beyond the limits of subatomic physics to include life and even consciousness. He stresses the existence of 'implicate' order, which exists 'folded up' in nature and gradually unfolds as the universe evolves, enabling organization to emerge.[10] One of the key features in quantum physics upon which Bohm draws in elaborating these ideas is non-locality, and it is to that topic which we now turn.

Non-locality in quantum mechanics

We have already seen how the results of an observation on an electron – which occupies a microscopic region of space – depend on the nature of a piece of macroscopic measuring apparatus – a coherently constructed entity organized over a large spatial region. What happens at a point in space thus depends intimately upon the wider environment, and in principle the whole universe. Physicists use the term *locality* to refer to situations where what happens at a point in space and time depends only on influences in the immediate vicinity of that point. Quantum mechanics is thus said to be 'non-local'.

Non-locality in quantum mechanics is most spectacularly manifested in certain situations generically known as EPR experiments after Einstein, Podolsky and Rosen, who first drew attention to the idea in the 1930s. Einstein was a persistent sceptic of quantum mechanics, and particularly disliked its non-locality because it seemed to bring quantum physics into conflict with his own theory of relativity.

He conceived of an experiment in which two particles interact and then separate to a great distance. Under these circumstances the quantum state of the combined system can be such that a measurement performed on one particle apparently affects the outcome of measurements performed on the other, distant, particle. This he found so unsettling he dubbed it 'spooky action-at-a-distance'.

More precisely, it is found that independently performed measurements made on widely separated particles yield correlated results.[11] This in itself is unsurprising because if the particles diverged from a common origin each will have retained an imprint of their encounter. The interesting point is the *degree* of correlation involved. This was investigated by John Bell of the CERN laboratory near Geneva.[12]

Bell showed that quantum mechanics predicts a significantly greater degree of correlation than can possibly be accounted for by any theory that treats the particles as independently real and subject to locality. It is almost as if the two particles engage in a conspiracy to cooperate when measurements are performed on them independently, even when these measurements are made simultaneously. The theory of relativity, however, forbids any sort of instant signalling or interaction to pass between the two particles. There seems to be a mystery, therefore, about how the conspiracy is established.

The conventional response to the EPR challenge was articulated by Niels Bohr. He argued that there is really no conflict with relativity after all if it is accepted that the two particles, although spatially separated, are still part of a unitary quantum system with a single wave function. If that is so, then it is simply not possible to separate the two particles physically, and to regard them as *independently real entities*, in spite of the fact that all forces acting directly between them are negligible over great distances. The independent reality of the particles comes only when measurements are performed on them. The mystery about how the particles conspire comes only if one insists on thinking about each of them possessing well-defined positions and motions prior to the observations.

The lesson of EPR is that quantum systems are fundamentally non-local. In principle, all particles that have ever interacted belong to a single wave function – a global wave function containing a stupendous number of correlations. One could even consider (and some physicists do) a wave function for the entire universe. In such a scheme the fate of any given particle is inseparably linked to the fate of the cosmos as a whole, not in the trivial sense that it may experience forces from its environment, but because its very reality is interwoven with that of the rest of the universe.

Quantum physics and life

Many of the physicists involved in developing quantum mechanics were fascinated by the implications of the new theory for biology. Some of them, such as Max Delbrück, made a career in biology. Others, including Bohr, Schrödinger, Pascual Jordan, Wigner and Elsasser, wrote extensively about the problems of understanding biological organisms from the physicist's point of view.

The Cosmic Blueprint

At first sight it might appear as if quantum mechanics is irrelevant to biology because living organisms are macroscopic entities. It must be remembered, however, that all the important processes of molecular biology are quantum in nature. Schrödinger showed that quantum mechanics is indispensable for understanding the stability of genetic information stored at the molecular level.

Accepting, then, that at the fundamental level life is encoded quantum mechanically, the question arises of how this quantum information is manifested in the form of a classical macroscopic organism. If heredity requires a quantum description, how does it relate to the purely classical concept of biological phenotype in interaction with the environment? This need to reconcile the quantum and classical descriptions of biological phenomena is really a version of the quantum measurement problem again.

Howard Pattee believes that the solution of the quantum measurement problem is intimately interwoven with the problem of understanding life. He points out that one of biology's essential characteristics, the transmission of hereditary information, requires a concept of *record*. And as we have seen, a quantum measurement takes place only when there is some sort of irreversible change leading to a trace or record.

Pattee refers to the level duality I have called 'hardware–software' as 'matter–symbol', and makes the provocative claim:[13] 'It is my central idea that the matter–symbol problem and the measurement or recording problem must appear at the origin of living matter.' He refers to enzymes as 'measuring molecules' and concludes that as no classical mechanism can provide the necessary speed and reliability for hereditary transmission 'life began with a catalytic coding process at the individual molecular level'.

If life is to be understood as an aspect of the quantum-classical reconciliation problem, then downward causation in biology would seem to be unavoidable. Furthermore, I believe we must then take seriously the non-local aspects of quantum physics in biological phenomena. As we have seen, the EPR experiment reveals how non-locality can manifest itself in correlations or 'conspiracy' over macroscopic distances. The two particles involved in that experiment are fundamentally inseparable in spite of their divergent locations; the system must be treated as a coherent whole. This is strongly reminiscent of biological processes.

There are many instances of biological phenomena where non-local effects seem to be at work. One of these is the famous protein folding problem. As mentioned in Chapter 7, proteins are formed as long chains which must then contort into a complicated and very specific three-dimensional shape before they can do their job properly. Biophysicists have long found the folding process enigmatic. How does the protein 'know' which final conformation to adopt?

It has been suggested that the required form is the most stable state (energy minimum), and hence the most probable state in some statistical sense. However, there are a great many other configurations with energies very nearly the same. If the protein had to explore all the likely possibilities before finding the right one it would take a very long time indeed. Somehow the protein seems to sense the needed final form and go for it. To achieve this action, widely separated portions of the protein have to move in unison according to an appropriate global schedule, otherwise the molecule would get tangled up in the wrong shape. This activity, which is a result of a plethora of quantum interactions, is clearly non-local in nature.

There are many other examples of this sort of 'action-at-a-distance', ranging from the fact that proteins bound at one site on a gene seem to exert an influence on other proteins bound thousands of atoms away, to the globally organized phenomenon of morphogenesis itself. There is, however, a fundamental difference between applying quantum mechanics to living organisms and its application to elementary particles. As pointed out by Niels Bohr, it is impossible to determine the quantum state of an organism without killing it. The irreducible disturbance entailed in making any sort of quantum measurement would totally disrupt the molecular processes so essential to life. Furthermore, it is not possible to compensate for this shortcoming by carrying out many partial measurements on a large collection of organisms, for all organisms are unique.

We here reach the central peculiarity concerning the application of quantum mechanics to any highly complex system. As explained, quantum mechanics is a statistical theory, and its predictions can only be verified by applying it to a collection of identical systems. This presents no problem in the case of elementary particles, which are inherently indistinguishable from other members of the same class (e.g. all electrons are alike). But when the system of interest is unique, a statistical prediction is irrelevant. This is certainly the case for a living organism, and must also be true for many complex inanimate systems, such as convection cells and Belousov–Zhabatinski patterns.

Elsasser has argued, in my view convincingly, that this uniqueness opens the way for the operation of additional organizing principles (his 'biotonic laws') that cannot be derived from the laws of quantum mechanics, yet do not contradict them:[14]

The primary laws are the laws of physics which can be studied quantitatively only in terms of their validity in homogeneous classes. There is then a 'secondary' type of order or regularity which arises only through the (usually cumulative) effect of individualities in inhomogeneous systems and classes. Note that the existence of such an order need not violate the laws of physics.

Elsasser recalls a proof by von Neumann that quantum mechanics cannot be supplemented by additional laws, but points out that in a sample of only one, the laws of quantum mechanics cannot be verified or falsified anyway. The proof is irrelevant. Quantum mechanics refers to the results of measurements on collections of identical systems, i.e. systems which belong to homogeneous classes. It has nothing to say, at least in its usual formulation, about regularities in *inhomogeneous* classes. But in biology one is interested in regularities in different but similar organisms, i.e. inhomogeneous classes. Quantum mechanics places no restriction on the existence of regularities of that sort. Therefore we are free to discover new, additional principles which refer to members of such classes. One such principle must surely be natural selection. It is hard to see how a description of natural selection could ever follow from the laws of quantum mechanics.

A rather different challenge to the applicability of quantum mechanics, in its present form, to biological systems comes from Robert Rosen, whose criticism of the narrow conceptual base of physics was discussed in Chapter 11. He maintains that some of the central assumptions which underlie the use of quantum mechanics in physics fail in biology. For example, when a physicist analyses a system quantum mechanically he first decides which dynamical quantities to use as 'observables', and constructs a mathematical formalism adapted to that choice. Typically the observables are familiar mechanical quantities – energy, position of a particle, spin, and so on. When it comes to biological systems, where interest lies with such concepts as mutation rate, enzyme recognition, DNA duplication, etc., it is far from clear what dynamical quantities these observables refer to.

More seriously, attempts to model biological activity at the molecular level in mechanical language encounter a deep conceptual obstacle. As remarked several times, the starting point of conventional mechanics is the construction of a Lagrangian for the system. The Lagrangian is closely related to another quantity, known as the Hamiltonian (after the Irish physicist William Rowan Hamilton). In classical mechanics, the Hamiltonian can be used to recover Newton's laws. Its importance, however, lies more with its role in quantum mechanics, for here we do not have Newton's laws as an alternative starting point. Quantization of a mechanical system *begins* with the Hamiltonian.

Rosen maintains that, in general, *no Hamiltonian exists* for biological systems. (Partly this is because they are open systems.) In other words, these systems cannot even be quantized by the conventional procedures. In fact, nobody really knows how to proceed when a system does not possess a Hamiltonian, so it is certainly premature to draw any conclusions about the application of quantum mechanics to biology – or any other complex system for which no Hamiltonian can be defined.

It is instructive to discover what the founding fathers of quantum mechanics thought about the validity of their theory when applied to biological phenomena. Schrödinger wrote[15] 'from all we have learnt about the structure of living matter, we must be prepared to find it working in a manner that cannot be reduced to the ordinary laws of physics'. But he is careful to explain that this is 'not on the ground that there is any "new force" or what not, directing the behaviour of the single atoms within a living organism'. It is, rather, because of the uniquely complex nature of living things, in which 'the construction is different from anything we have yet tested in the physical laboratory'.

In his biography of Schrödinger, physicist William Scott discusses his interpretation of the great man's position concerning new organizing principles in the higher-level sciences such as biology:[16]

In the light of the above analysis, it appears that Schrödinger's claim about new laws of physics and chemistry which may appear in biology is largely a matter of terminology. If the terms 'physics and chemistry' are to keep their present meaning, Schrödinger's prediction should be interpreted to mean that new organizing principles will be found that go beyond the laws of physics and chemistry but are not in contradiction to these laws.

The additional freedom for such new organizing principles to act comes, asserts Scott, from 'the range of possible initial or boundary values. In systems as complex as living organisms, this range of freedom is very great indeed'.

Niels Bohr thought deeply about the nature of living organisms, which he insisted were primary phenomena that could not be reduced to atomic-level activity:[17]

On this view, the existence of life must be considered an elementary fact that cannot be explained, but must be taken as a starting point in biology.

Werner Heisenberg describes a conversation with Bohr during a boat trip in the early thirties.[13] Heisenberg expressed reservations about the adequacy of quantum mechanics to explain biology. He asked Bohr whether he believed that a future unified science that could account for biological phenomena would simply consist of quantum mechanics plus some biological concepts superimposed, or whether – 'this unified science will be governed by broader natural laws of which quantum mechanics is only a limiting case'.

Bohr, in his typically elder statesman manner, dismissed the relevance of the distinction, preferring to fall back on his famous 'principle of complementarity'. Biological and physical descriptions are, he asserted, merely two complementary rather than contradictory ways of looking at nature. But what about evolution, pressed Heisenberg. 'It is very difficult to believe that such complicated organs as, for instance, the human eye were built up quite gradually as the result of purely accidental changes.' Bohr conceded that the idea of new forms originating through pure accident 'is much more questionable, even though we can hardly conceive of an alternative'. Nevertheless, he preferred to 'suspend judgement'.

Heisenberg finally tackled Bohr on the issue of consciousness. Did not the existence of consciousness attest to the need for an extension to quantum theory? Bohr replied that this argument 'looks highly convincing at first sight . . . consciousness must be a part of nature . . . which means that, quite apart from the laws of physics and chemistry, as laid down in quantum theory, we must also consider laws of a quite different kind'.

What are we to conclude from all this?

The laws of quantum mechanics are not themselves capable of explaining life, yet they do open the way for the operation of non-local correlations, downward causation and new organizing principles. It may be that such principles could remain consistent with quantum mechanics, or it may be that quantum mechanics fails above a certain level of organizational complexity. Whatever the case, it is clearly a gross error to envisage biological organisms as classical machines, operating solely by the rearrangement of molecular units subject only to local forces. And this error becomes all the more forcefully apparent when the existence of consciousness is considered.

13
Mind and Brain

Patterns that think

In any discussion of complexity and self-organization, the brain occupies a special place, for once again we cross a threshold to a higher conceptual level. We now enter the world of *behaviour*, and eventually of consciousness, free will, thoughts, dreams, etc. It is a field where subjective and objective become interwoven, and where deep-seated feelings and beliefs inevitably intrude. It is probably for that reason that physical scientists seem to avoid discussing the subject. But sooner or later the question has to be addressed of whether *mental* functions can ultimately be reduced to physical processes in the brain, and thence to physics and chemistry, or whether there are additional laws and principles pertaining to the mental world that cannot be derived mechanistically from the physics of inanimate matter.

From the viewpoint of neurophysiology, the brain can be studied at two levels. The lower level concerns the workings of individual neurones (brain cells) and their interconnections, establishing what makes them fire and how the electrical pulses propagate between neurones. At a higher level, the brain can be regarded as a fantastically complex *network* around which electrical *patterns* meander. If, as seems clear, mental processes are associated with patterns of neural activity rather than the state of any particular neurone, then it is the latter approach that is most likely to illuminate the higher functions of behaviour and consciousness.

Many attempts have been made to model neural nets after the fashion of cellular automata, by adopting some sort of wiring system together with a rule for evolving the electrical state of the net deterministically in time, and then running a computer simulation. These studies are motivated in part by practical considerations. Computer designers are anxious to discover how the brain performs certain integrative tasks, such as pattern recognition, so as to be able to design 'intelligent' machines. There is also a desire to find simple models that might give a clue to certain basic mental functions, such as dreaming, memory storage and recall, as well as malfunctions like epileptic seizures.

Neural anatomy is awesomely complex. The human brain contains some hundred billion neurones and a given neurone may be directly connected to a great many others. It seems probable that some of the interconnections are structured systematically, while others are random. The electrical output signal of a given neurone will depend in a non-linear way on the combined input it receives from its connected partners. These inputs may have both an excitatory and an inhibitory effect. Thus, the character of the output signal from a particular neurone, such as the rate of firing, depends in a very complex way on what is happening elsewhere in the system.

It is no surprise that a system with such a high degree of non-linearity and feedback should display self-organizing capabilities and evolve collective modes of behaviour leading to the establishment of global patterns and cycles. The trick is to capture some of these features in a tractable computational model.

A typical model network might consist of a few hundred elements ('neurones') each randomly connected to about 20 other elements with various different strengths. The neurones are attributed a specified recovery time between subsequent firings. The system is then put in an initial state by specifying a particular firing pattern, perhaps at random, and then evolving deterministically – by computer simulation – to see what patterns establish themselves.

An important refinement is to introduce *plasticity* into the system, by continually modifying the network parameters until interesting behaviour is encountered. It is thought that the brain employs plasticity in its own development. In one model, formulated at Washington University, St Louis, the net is modified with time in a way that depends on the current neuronal activity: the interconnection strengths are changed according to whether the end neurones are firing or not. This inter-level feedback enables the net to evolve some remarkable capabilities. One plasticity algorithm, known as *brainwashing*, systematically weakens the connections between active neurones, thereby reducing the level of activity. The result is that instead of the net engaging in uninteresting high-level activity, it displays self-organizing behaviour, typically settling into cyclic modes of various periods and complexities. The St Louis team believe that their model offers a plausible first step towards a network capable of learning, and ultimately displaying intelligent behaviour.

Memory

A major contribution to the study of neural nets was made in 1982 by J. J. Hopfield of Caltech. In Hopfield's model each neurone can be in one of only two states, firing or quiescent. Which state it is in is decided by whether its electric potential exceeds a certain threshold, this being determined by the level of incoming signals from its connecting partners. The level is in turn dependent on the strengths of the various interconnections. The model assumes the connection strengths are symmetric, i.e. *A* couples to *B* as strongly as *B* couples to *A*. Of course, the coupling strengths are only relevant when the neurones on the ends of the connection are active. One can choose both positive (excitatory) and negative (inhibitory) strengths.

The attractiveness of the Hopfield model is that is possesses a readily visualizable physical analogue. The various possible states of the network can be represented by a bumpy surface in space, with the current state corresponding to an imaginary ball rolling on the surface. The ball will tend to roll down into the valleys, or basins of attraction, seeking out the local minima. This tells us that the net will tend to settle into those firing patterns corresponding to 'minimum potential' states. The height of the surface at each point can be envisaged as analogous to energy, and is determined by the combined strengths of the interconnections: large positive strengths contribute small energies. The favoured 'valley' states are therefore those in which strongly connected neurones tend to fire together. The model may be studied by (metaphorically) manipulating the energy landscape and searching for interesting behaviour.

Further realism can be added by including the analogue of thermal noise. If the ball were to be continually jiggled around, it would have the opportunity to explore the landscape more thoroughly. For example, there would be a chance that it would vacate one valley and find another deeper valley nearby. On average, it would spend its time near the deepest minima. To effect this refinement, it is merely necessary to introduce a random element into the firing rule. Using such probabilistic nets, rudimentary learning and recognition functions have been observed.

It has been conjectured that the Hopfield model provides an important form of memory. The idea is that the neural activity pattern corresponding to the base of a valley represents some concept or stored information. To access it, one simply places the imaginary ball somewhere in the basin and waits for it to roll down to the bottom. That is, the firing pattern needs only to be rather close to that representing the target concept for the activity to evolve towards it. The net then repeats the relevant pattern of activity, and so reproduces the stored information. This facility is called content-addressable memory by computer scientists, because it enables a complete concept to be recovered from a fragment. It corresponds to what happens when we 'search our brains' for an idea or memory based on some vague recollection or associated image.

The Hopfield model of memory differs fundamentally from that used in computers, where each bit of information is stored on a specific element and can only be recalled by specifying the exact address. In the Hopfield case the information is stored *holistically*; it is the collective pattern of activity throughout the net that represents the information. Whereas computer memories are vulnerable to single element failure, the Hopfield system is highly robust, because the functioning of the neural activity does not depend crucially on any particular neurone. Clearly something like this must happen in the brain, where neurones frequently die without noticeably inhibiting the brain's functioning.

The concept of learning can be captured in this model by envisaging that the imaginary landscape can be remodelled by external input, creating new valleys representing freshly stored information. This involves plasticity of the sort already described. Hopfield has also found that memory access works more efficiently if an 'unlearning' process is included – like a learning algorithm, but reversed – and has even conjectured that something like this may be going on in the brain during periods of dreaming sleep.[1]

These recent exciting advances in modelling neural networks emphasize the importance of the collective and holistic properties of the brain. They show that what matters is the *pattern* of neural activity, not the detailed functioning of individual neurones. It is at this collective level that new qualities of self-organization appear, which seem to have their own rules of behaviour that cannot be derived from the laws of physics governing the neural function. Indeed, in the computer simulations I have discussed here, there *is* no physics involved, except in so far as realistic firing procedures are specified.

Behaviour

Whatever the mechanisms by which the brain functions, the result in the real world is that organisms possessed of them – or even rudimentary nervous systems – display *complex behaviour*. Behaviour represents a new and still higher level of activity in nature. If organic functions prove hard to reduce to physics, behaviour is well-nigh impossible.

Consider, for example, a dog following the scent of a quarry. The organism as a whole operates to carry out a specific task as an integrated entity. The task involves an enormously complex collection of interlocking functions, all of which must be subordinated to the overall strategy. It is almost impossible to resist the impression that a dog following a scented trail is acting *purposefully*, with some sort of internal predictive model of the final state it is attempting to achieve; in this case, seizing the quarry.

A complete reductionist would be hard put to explain the dog's strongly teleological behaviour. Each atom of the dog is supposed to move in accordance only with the blind forces acting upon it by neighbouring atoms, all of which are simply following the dictates of the laws of physics. Yet who can deny that the dog is somehow manipulating its body towards the seizure of the quarry?

It is important not to fall into the trap of supposing that all purposeful behaviour is consciously considered. A spider weaving a web, or a collection of ants building a nest surely have no conscious awareness of what they are doing (at least, they have no conception of the overall strategy), and yet they still accomplish the task. The whole domain of instinct falls into this category. According to standard theory, the remarkable instinctive abilities of insects and birds is entirely due to genetic programming. In other words, nobody teaches a spider how to weave a web; it inherits the skill through its DNA.

Of course, nobody has the slightest idea of how the mere fact of arranging a few molecules in a particular permutation (a static form) brings about highly integrated *activity*. The problem is far worse here than in morphogenesis, where spatial patterns are the end product. It might be conjectured that the genetic record resembles a sequence of programmed instructions to be 'run', like the punched-tape input of a pianola, but this analogy doesn't stand up to close scrutiny. Even instinctive behavioural tasks can be disrupted without catastrophic consequences. An obstacle placed in an ant trail may cause momentary confusion, but the ants soon establish an adjusted strategy to accommodate the new circumstances.

Obviously there are a host of control and compensating mechanisms that depend on sensory input for their operation, and not on the mechanistic implementation of a fixed set of instructions. In other words, the organism cannot be regarded (like the pianola) as a closed system with a completely determined repertoire of activity. An ant must be seen as part of a colony and the colony as part of the environment. The concept of ant behaviour is thus holistic, and only partially dependent on the internal genetic make-up of an individual ant.

Perhaps the most striking examples of the robustness of instinctive behaviour come from bird migration experiments. It is well known that birds can perform fantastic feats of navigation, for which purpose they are apparently assisted by astronomy and the Earth's magnetism. Some birds fly for thousands of miles with pinpoint precision, in spite of the fact that they are never taught an itinerary. Most remarkable of all are those cases where birds are taken hundreds or even thousands of miles from home, to a part of the Earth of which they can have no knowledge, and on release fly back in virtually a straight line.

Again, the conventional response to these astonishing accomplishments is to suppose that navigational skills are genetically programmed, i.e. stored in the birds' DNA. But in the absence of an explanation for how an arrangement of molecules translates into a behavioural skill that can accommodate completely unforeseen disruption, this is little more than hand-waving.

If the necessary astro-navigational information is built into the DNA molecule, it implies that in principle, given a sufficient understanding of the nature of DNA, one could 'decode' this information and reconstruct a map of the stars! More than this. The bird needs to know times as well as orientations, so the astronomical panorama would actually be a movie. Letting one's imagination have free reign, one cannot help but wonder whether a clever scientist who had never seen either a bird or the sky could, by close examination of a single molecule of DNA, figure out the details of a rudimentary planetarium show!

It seems to me far more plausible that the secret of the bird's navigational abilities lies in an altogether different direction. As we have seen, it is a general property of complex systems that above a certain threshold of complexity, new qualities emerge that are not only absent, but simply meaningless at a lower conceptual level. At each transition to a higher level of organization and complexity new laws and principles must be invoked, in addition to the underlying lower level laws, which may still remain valid (or they may not, of course).

When it comes to animal behaviour, the relevant concepts are informational in character (the bird *navigates* according to *star positions*) so one expects laws and principles that refer to the quality, manipulation and storage of information to be relevant – the sort of thing hinted at in the study of cellular automata and neural nets. Such laws and principles cannot be reduced to mechanistic physics, a subject which is simply irrelevant to the phenomenon.

Consciousness

The teleological quality of behaviour becomes impossible to deny when it is consciously pursued, for we know from direct experience that we often *do* have a preconceived image of a desired end state to which we strive. When we enter the realm of conscious experience, we again cross a threshold of organizational complexity that throws up its own new concepts – thoughts, feelings, hopes, fears, memories, plans, volitions. A major problem is to understand how these *mental events* are consistent with the laws and principles of the physical universe that produces them.

The reductionist is here presented with a severe difficulty. If neural processes are nothing but the motions of atoms and electrons slavishly obeying the laws of physics, then mental events must be denied any distinctive reality altogether, for the reductionist draws no fundamental distinction between the physics of atoms and electrons in the brain and the physics of atoms and electrons elsewhere. This certainly solves the problem of the consistency between the mental and physical world.

However, one problem is solved only to create another. If mental events are denied reality, reducing humans to mere automata, then the very reasoning processes whereby the reductionist's position is expounded are also denied reality. The argument therefore collapses amid its own self-reference.

On the other hand, the assumption that mental events are real is not without difficulty. If mental events are in some way *produced* by physical processes such as neural activity, can they possess their own independent dynamics?

The difficulty is most acutely encountered in connection with volition, which is perhaps the most familiar example of downward causation. If I decide to lift my arm, and my arm subsequently rises, it is natural for me to suppose that my *will* has *caused* the movement. Of course, my mind does not act on my arm directly, but through the intermediary of my brain. Evidently the act of my willing my arm to move is associated with a change in the neural activity of my brain – certain neurones are 'triggered' and so forth – which sets up a chain of signals that travel to my arm muscles and bring about the required movement.

There is no doubt that this phenomenon – part of what is known as the *mind-body problem* – presents the greatest difficulty for science. On the one hand, neural activity in the brain is supposed to be determined by the laws of physics, as is the case with any electrical network. On the other hand, direct experience encourages us to believe that, at least in the case of intended action, that action is caused by our mental states. How can one set of events have two causes?

Opinions on this issue range from the above-mentioned denial of mental events, called behaviourism, to idealism, in which the *physical* world is denied and all events are regarded as mental constructs. Undoubtedly relevant to this issue is the fact that the brain is a highly nonlinear system and so subject to chaotic behaviour. The fundamental upredictability of chaotic systems and their extreme sensitivity to initial conditions endows them with an open, whimsical quality. Physicist James Crutchfield and his colleagues believe that chaos provides for free will in an apparently deterministic universe:[2]

> Innate creativity may have an underlying chaotic process that selectively amplifies small fluctuations and molds them into macroscopic coherent mental states that are experienced as thoughts. In some cases the thoughts may be decisions, or what are perceived to be the exercise of will. In this light, chaos provides a mechanism that allows for free will within a world governed by deterministic laws.

Among the many other theories of mind-body association are Cartesian dualism, whereby an external independently existing mind or soul exerts mystical forces on the brain to induce it to comply with the will. Then there is psychophysical parallelism, which admits mental events, but ties them totally to the physical events of the brain and denies them any causal potency. There is also something called functionalism, which draws analogies between mental events and computer software. Yet another idea is panpsychism, which attributes a form of consciousness to everything. This has been espoused by Teilhard de Chardin, and more recently by the physicist Freeman Dyson, who writes:[3]

> I think our consciousness is not just a passive epiphenomenon carried along by the chemical events in our brains, but is an active agent forcing the molecular complexes to make choices between one quantum state and another. In other words, mind is already inherent in every electron. . .

Mind and Brain

I do not wish to review these many and contentious theories here. My concern is to affirm the reality of mental events and to show how they comply with the central theme of this book – that each new level of organization and complexity in nature demands its own laws and principles.

For this purpose I have been much inspired by the work of the Nobel prizewinner R. W. Sperry who has conducted some fascinating experiments on 'split brain' subjects. These are patients who have had the left and right hemispheres of their brains surgically disconnected for medical reasons. As a result of his experiments, Sperry eschews reductionist explanations of mental phenomena, and argues instead for the existence of something like downward causation (it is technically known as emergent interactionism).

Sperry regards mental events as[4] 'holistic configurational properties that have yet to be discovered' but which will turn out to be 'different from and more than the neural events of which they are composed . . . they are emergents of these events'. He subscribes to the idea that higher-level entities possess laws and principles in their own right that cannot be reduced to lower-level laws:

> These large cerebral events as entities have their own dynamics and associated properties that causally determine their interactions. These top-level systems' properties supersede those of the various subsystems they embody.

Thus mental events are ascribed definite causal potency; they can make things happen.

How, then, does Sperry explain the peaceful coexistence of top and bottom level laws, one set controlling the neural patterns (holistic configurational properties) and another the atoms of which the neurones are composed? He explicitly states that[5] 'mental forces or properties exert a regulative control influence in brain physiology'. In other words, mind (or the collective pattern of neuronal activity) somehow produces forces that act on matter (the neurones). Nevertheless, Sperry is at pains to point out that this example of downward causation in no way violates the lower-level laws.

How is this achieved?[6]

> The way in which mental phenomena are conceived to control the brain's physiology can be understood very simply in terms of the chain of command of the brain's hierarchy of causal controls. It is easy to see that the forces

operating at subatomic and subnuclear levels within brain cells are molecule-bound, and are superseded by the encompassing configurational properties of the brain molecules in which the subatomic elements are embedded.

Sperry talks of the lower-level entities becoming 'caught up' in the holistic pattern, much as a water droplet is caught up in a whirlpool and constrained to contribute cooperatively to the overall organized activity.

Central to Sperry's position is that the agency of causation can be different at different levels of complexity, and that moreover, causation can operate simultaneously at different levels, and between levels, without conflict. Thus thoughts may cause other thoughts, and the movement of electrons in the brain may cause other electrons to move. But the latter is not a complete explanation of the former, even though it is an essential element of it:[7]

> Conscious phenomena [are] emergent functional properties of brain processing [which] exert an active control role as causal determinants in shaping the flow patterns of cerebral excitation. Once generated from neural events, the higher order mental patterns and programs have their own subjective qualities and progress, operate and interact by their own causal laws and principles which are different from, and cannot be reduced to those of neurophysiology ... The mental forces do not violate, disturb, or intervene in neuronal activities but they do supervene... Multilevel and interlevel causation is emphasized in addition to the one-level sequential causation traditionally dealt with.

Although physicists tend to react to such ideas with horror, they seem to be perfectly acceptable to computer scientists, artificial intelligence experts and neuroscientists. Donald MacKay, Professor of Communication and Neuroscience at the University of Keele, also accepts that causation can operate differently at different levels. He points out that there has been[8]

> an expansion of our concepts of causality that has come with developments in the theory of information and control. In an information system, we can recognize 'informational' causality as something qualitatively distinct from physical causality, coexisting with the latter and just as efficacious. Roughly speaking, whereas in classical physics the determination of force by force requires a flow of energy, from the standpoint of information theory the determination of form by form requires a flow of information. The two are so different that a flow of information from A to B may require a flow of energy from B to A; yet they are totally interdependent and complementary, the one process being embodied in the other.

Mind and Brain

American artificial intelligence researcher Marvin Minsky, writes:[9]

Many scientists look on chemistry and physics as ideal models of what psychology should be like. After all, the atoms in the brain are subject to the same all-inclusive laws that govern every other form of matter. Then can we also explain what our brains actually do entirely in terms of those same basic principles? The answer is no, simply because even if we understand how each of our billions of brain cells work separately, this would not tell us how the brain works as an agency. The 'laws of thought' depend not only upon the properties of those brain cells, but also on how they are connected. And these connections are established not by the basic, 'general' laws of physics, but by the particular arrangements of the millions of bits of information in our inherited genes. To be sure, 'general' laws apply to everything. But, for that very reason, they can rarely explain anything in particular.

Does this mean that psychology must reject the laws of physics and find its own? Of course not. It is not a matter of *different* laws, but of *additional* kinds of theories and principles that operate at higher levels of organization.

Minsky makes the important point that the brain, as is the case with a computer, is a *constrained* system. The permissible dynamical activity is dependent both on the laws of physics *and* on the 'wiring' arrangement. It is the presence of constraints – which cannot themselves be deduced from the laws of physics because they refer to *particular* systems – that enable new laws and principles to be realized at the higher level. Thus a computer may be programmed to play chess or some other game on a screen. The rules of the game determine the 'laws' whereby the images move about on the screen; that is, they fix a rudimentary dynamics of the higher-level entities (the 'chess pieces'). But there is, of course, no conflict between the laws of chess obeyed by the images and the underlying laws of physics ultimately controlling the electrons in the circuitry and impinging on the screen.

These and other considerations have convinced me that there are new processes, laws and principles which come into play at the threshold of mental activity. I do not believe that behaviour, let alone psychology, can ultimately be reduced to particle physics. I find it absurd to suppose that the migratory habits of birds, not to mention my personal sensations and emotions, are all somehow contained in the fundamental Lagrangian of superstring theory or whatever.

I also contend that we will never fully understand the lower level processes until we also understand the higher level laws. Problems such as the collapse of the wave function in quantum mechanics, which affect the very consistency of particle physics, seem to demand the inclusion of the observer in a fundamental way. I think that observation in quantum mechanics must ultimately be referred to the upper level laws that govern the mental events to which the act of observation couples the microscopic events.

I finish this section with a quote from the physical chemist Michael Polanyi, which expresses similar sentiments:[10]

> There is evidence of irreducible principles, additional to those of morpholo-gical mechanisms, in the sentience that we ourselves experience and that we observe indirectly in higher animals. Most biologists set aside these matters as unprofitable considerations. But again, once it is recognized, on other grounds, that life transcends physics and chemistry, there is no reason for suspending recognition to the obvious fact that consciousness is a principle that fundamentally transcends not only physics and chemistry but also the mechanistic principles of living beings.

Beyond consciousness

Mental events do not represent the pinnacle of organization and complexity in nature. There is a further threshold to cross yet, into the world of culture, social institutions, works of art, religion, scientific theories, literature, and the like. These abstract entities transcend the mental experiences of individuals and represent the collective achieve-ments of human society as a whole. They have been termed by Popper 'World 3' entities – those of World 1 being material objects, and those of World 2 being mental events.

It is important to appreciate that the existence of social organization – which carries with it its own irreducible laws and principles – is not dependent upon mental events. Many insects have elaborate societies, presumably without being remotely aware of the fact. Human society, however, whatever its biological origins, has evolved to the stage where it is shaped and directed by conscious decisions, and this has produced World 3.

Can World 3 be reduced to World 2, or even World 1? I do not see how this can be the case, for World 3 entities possess logical and structural relationships of their own that transcend the properties of individual human beings. Take mathematics, for example. The prop-erties of real numbers amount to far more than our collective experiences of arithmetic. There will be theorems concerning numbers which are unknown to anybody alive today, yet which are nevertheless true. In music, a concerto has its own internal organization and consistency independently of whether anybody is actually listening to it being played. Moreover, some World 3 entities, such as criminal data banks or money market records, are completely beyond the capacity of any one individual to know, yet they still exist.

World 3 systems have their own dynamical behaviour. The principles of economics, rough though they may be, cannot be reduced to the laws of physics. World 3 entities clearly have causal potency of their own. A stock market crash may legitimately be attributed to a change of government – another World 3 event. It is hard to see how a causal link of this sort could ever be recovered from the causal processes of atoms.

Again we find many examples of downward causation, wherein World 3 entities can be considered responsible for bringing about changes in Worlds 2 and 1. Thus an artistic tradition might inspire a sculptor to shape a rock into a particular form. The thoughts of the sculptor, and the distribution of atoms in the rock, are here determined by the abstract World 3 entity 'artistic tradition'. Similarly, a new mathematical theorem or scientific theory may lead a scientist to conduct a previously unforeseen experiment.

With World 3 we also reach the end of the chain of interaction discussed in Chapter 12 in connection with the quantum measurement problem, for it is here that we arrive at the concept of *meaning*. Wheeler uses the definition of the Norwegian philosopher D. Follesdal: meaning is the joint product of all the evidence available to those who communicate. It is therefore a collective, cultural attribute. Indeed, we must regard any sort of scientific measurement as part of a cultural enterprise, for it is always conducted within the context of a scientific theory, or at least a conceptual framework, derived from the community as a whole.

Starting with the fundamental subatomic entities, we have explored the progression of organization and complexity upwards – through inanimate states of matter, living organisms, brains, minds and social systems – to World 3. Is this the end of the upward ladder? Does anything lie beyond?

Many people, of course, believe that something does lie beyond. Those of a religious persuasion see man and his culture as a relatively low-level manifestation of reality. Some conjecture about higher levels of organizational power, and even downward causation from 'above' shaping the events of Worlds 1, 2 and 3. In this context it is possible to regard nature itself, including its laws, as an expression of a higher organizing principle.

On a less cosmic level, there are still many beliefs that place the individual mind only part-way up the organizational ladder. Jung's theory of the collective unconscious, for example, treats the individual mind as only one component in a shared cultural experience from which it may draw. Mystical ideas like astrology likewise regard individual minds as subordinated to a global harmony and organization that is reflected in astronomical events. Those who believe in fate or destiny must also require a higher organizational principle that moulds human experiences in accordance with some teleological imperative.

Lastly, there are a great many people who appeal to the sort of ideas I have been expounding in this book to justify their belief in the so-called paranormal. They regard alleged phenomena such as extra-sensory perception, telepathy, precognition and psychokinesis as evidence of organizational principles that extend beyond the individual mind, and allow for the downward causation of mind over matter, often in flagrant violation of the laws of physics.

All I wish to say on this score is that it is one thing to expose the limitations of reductionism; it is quite another to use those limitations for an 'anything now goes' policy. Perhaps one day paranormal phenomena will become normal, or maybe they will finally be discounted as groundless. Whatever is the case, the decision must be based on sound scientific criteria, and not just a sweeping rejection of an uncomfortable paradigm.

Leaving aside these religious or speculative ideas, there is still a sense in which human mind and society may represent only an intermediate stage on the ladder of organizational progress in the cosmos. To borrow a phrase from Louise Young, the universe is as yet 'unfinished'. We find ourselves living at an epoch only a few billion years after the creation. From what can be deduced about astronomical processes, the universe could remain fit for habitation for trillions of years, possibly for ever. The heat death of the cosmos, a concept that has dogged us throughout, poses no threat in the imaginable future, and by the time scale of human standards it is an eternity away.

As our World 3 products become ever more elaborate and complex (one need only think of computing systems) so the possibility arises that a new threshold of complexity may be crossed, unleashing a still higher organizational level, with new qualities and laws of its own. There may emerge collective activity of an abstract nature that we can scarcely imagine, and may even be beyond our ability to conceptualize. It might even be that this threshold has been crossed elsewhere in the universe already, and that we do not recognize it for what it is.

14
Is there a Blueprint?

Optimists and pessimists

Most scientists who work on fundamental problems are deeply awed by the subtlety and beauty of nature. But not all of them arrive at the same interpretation of nature. While some are inspired to believe that there must be a meaning behind existence, others regard the universe as utterly pointless.

Science itself cannot reveal whether there is a meaning to life and the universe, but scientific paradigms can exercise a strong influence on prevailing thought. In this book I have sketched the story of a new, emerging paradigm that promises to radically transform the way we think about the universe and our own place within it. I am convinced that the new paradigm paints a much more optimistic picture for those who seek a meaning to existence. Doubtless there will still be pessimists who will find nothing in the new developments to alter their belief in the pointlessness of the universe, but they must at least acknowledge that the new way of thinking about the world is more cheerful.

The theme I have been presenting is that science has been dominated for several centuries by the Newtonian paradigm which treats the universe as a mechanism, ultimately reducible to the behaviour of individual particles under the control of deterministic forces. According to this view, time is merely a parameter; there is no real change or evolution, only the rearrangement of particles. The laws of thermodynamics reintroduced the notion of flux or change, but the reconciliation of the Newtonian and thermodynamic paradigms led only to the second law, which insists that all change is part of the inexorable decay and degeneration of the cosmos, culminating in a heat death.

The emerging paradigm, by contrast, recognizes that the collective and holistic properties of physical systems can display new and unforeseen

197

modes of behaviour that are not captured by the Newtonian and thermodynamic approaches. There arises the possibility of *self-organization*, in which systems suddenly and spontaneously leap into more elaborate forms. These forms are characterized by greater complexity, by cooperative behaviour and global coherence, by the appearance of spatial patterns and temporal rhythms, and by the general unpredictability of their final forms.

The new states of matter require a new vocabulary, which includes terms like growth and adaptation – concepts more suited to biology than physics or chemistry. There is thus a hint of unification here. Above all, the new paradigm transforms our view of time. Physical systems can display unidirectional change in the direction of *progress* rather than decay. The universe is revealed in a new, more inspiring light, unfolding from its primitive beginnings and progressing step by step to ever more elaborate and complex states.

The resurgence of holism

Many non-scientists find both the Newtonian and thermodynamic paradigms profoundly depressing. They use reductionism as a term of abuse. They regard its successes as somehow devaluing nature, and when applied in the life sciences, devaluing themselves. In a recent television debate in which I took part, the audience was invited to express views about science and God. An irate man complained bitterly. 'Scientists claim that when I say to my wife "I love you" that is nothing but one meaningless mound of atoms interacting with another meaningless mound of atoms.' Such despair over the perceived sterility of reductionist thinking has led many people to turn to holism. In this, they have no doubt been greatly encouraged by the recent resurgence of holistic thinking, in sociology, medicine and the physical sciences.

Yet it would be a grave mistake to present reductionism and holism as somehow locked in irreconcilable combat for our allegiance. They are really two complementary rather than conflicting paradigms. There has always been a place for both in properly conducted science, and it is a gross simplification to regard either of them as 'right' or 'wrong'.

Those who would appeal to holism must distinguish between two claims. The first is the statement that as matter and energy reach higher, more complex, states so new qualities *emerge* that can never be embraced by a lower-level description. Often cited are life and consciousness, which are simply meaningless at the level of, say, atoms.

Examples of this sort seem to be, quite simply, incontrovertible facts of existence. Holism in this form can only be rejected by denying the reality of the higher-level qualities, e.g. by claiming that consciousness does not really exist, or by denying the meaningfulness of higher-level concepts, such as a biological organism. Since I believe that it is the job of science to explain the world as it appears to us, and since this world includes such entities as bacteria, dogs and humans, with their own distinctive properties, it seems to me at best evasive, at worst fraudulent, to claim that these properties are explained by merely defining them away.

More controversial, however, is the claim that these higher-level qualities demand higher-level laws to explain them. We met this claim, for example, in the suggestion that there exist definite *biotonic* laws for organic systems, and in the ideas of dialectical materialism, which holds that each new level in the development of matter brings its own laws that cannot be reduced to those at lower levels. More generally we saw the possibility of three different types of organizing principles: weak, strong and logical.

The existence of logical organizing principles seems to be fairly well established already, for example, in connection with chaotic systems and Feigenbaum's numbers. Weak organizing principles, in the form of the need to specify various boundary conditions and global constraints are accepted at least as a methodological convenience.

Strong organizing principles – additional laws of physics that refer to the cooperative, collective properties of complex systems, and which cannot be derived from the underlying existing physical laws – remain a challenging but speculative idea. Mysteries such as the origin of life and the progressive nature of evolution encourage the feeling that there are additional principles at work which somehow make it 'easier' for systems to discover complex organized states. But the reductionist methodology of most scientific investigations makes it likely that such principles, if they exist, risk being overlooked in current research.

Predestiny

The new paradigm will drastically alter the way we view the evolution of the universe. In the Newtonian paradigm the universe is a clockwork, a slave of deterministic forces trapped irretrievably on a predetermined pathway to an unalterable fate. The thermodynamic paradigm gives us a universe that has to be started in an unusual state of order, and then degenerates. Its fate is equally inevitable, and uniformly bad.

In both the above pictures *creation* is an instantaneous affair. After the initial event nothing fundamentally new ever comes into existence. In the Newtonian universe atoms merely rearrange themselves, while in the thermodynamic picture the history of the universe is one of *loss*, leading towards dreary featurelessness.

The emerging picture of cosmological development is altogether less gloomy. Creation is not instantaneous; it is an ongoing process. The universe has a life history. Instead of sliding into featurelessness, it rises out of featurelessness, growing rather than dying, developing new structures, processes and potentialities all the time, unfolding like a flower.

The flower analogy suggests the idea of a blueprint – a pre-existing plan or project which the universe is realizing as it develops. This is Aristotle's ancient teleological picture of the cosmos. Is it to be resurrected by the new paradigm of modern physics?

It is important to appreciate that according to the new paradigm determinism is irrelevant: the universe is intrinsically unpredictable. It has, as it were, a certain 'freedom of choice' that is quite alien to the conventional world view. Circumstances constantly arise in which many possible pathways of development are permitted by the bottom-level laws of physics. Thus there arises an element of novelty and creativity, but also of uncertainty.

This may seem like cosmic anarchy. Some people are happy to leave it that way, to let the universe explore its potentialities unhindered. A more satisfactory picture, however, might be to suppose that the 'choices' occur at critical points (mathematicians would call them singularities in the evolution equations) where new principles are free to come into play, encouraging the development of ever more organized and complex states. In this more *canalized* picture, matter and energy have innate self-organizing tendencies that bring into being new structures and systems with unusual efficiency. Again and again we have seen examples of how organized behaviour has emerged unexpectedly and spontaneously from unpromising beginnings. In physics, chemistry, astronomy, geology, biology, computing – indeed, in every branch of science – the same propensity for self-organization is apparent.

The latter philosophy has been called 'predestinist' by the biologist Robert Shapiro, because it assumes that the present form and arrangement of things is an inevitable outcome of the operation of the laws of nature. I suspect he uses the term pejoratively, and I dislike the mystical flavour it conveys. I prefer the word *predisposition*.

Is there a Blueprint?

Who are the predestinists?

Generally speaking, they are those who are not prepared to accept that certain key features of the world are simply 'accidents' or quirks of nature. Thus, the existence of living organisms does not surprise a predestinist, who believes that the laws of nature are such that matter will inevitably be led along the road of increasing complexity towards life. In the same vein, the existence of intelligence and conscious beings is also regarded as part of a natural progression that is somehow built into the laws. Nor is it a surprise to a predestinist that life arose on Earth such a short period of time (geologically speaking) after our planet became habitable. It would do so on any other suitable planet. The ambitious programme to search for intelligent life in space, so aptly popularized by Carl Sagan, has a strong predestinist flavour.

Predestiny – or predisposition – must not be confused with predeterminism. It is entirely possible that the properties of matter are such that it does indeed have a propensity to self-organize as far as life, given the right conditions. This is not to say, however, that any particular life form is inevitable. In other words, predeterminism (of the old Newtonian sort) held that everything *in detail* was laid down from time immemorial. Predestiny merely says that nature has a predisposition to progress along the general lines it has. It therefore leaves open the essential unknowability of the future, the possibility for real creativity and endless novelty. In particular it leaves room for human free will.

The belief that the universe has a predisposition to throw up certain forms and structures has become very fashionable among cosmologists, who dislike the idea of special initial conditions. There have been many attempts to argue that something close to the existing large-scale structure of the universe is the inevitable consequence of the laws of physics whatever the initial conditions. The inflationary universe scenario is one such attempt. Another is Penrose's suggestion that the initial state of the universe follows from some as-yet unknown physical principle. A third is the attempt by Hawking and co-workers to construct a mathematical prescription that will fix in a 'natural' way the quantum state of the universe.[1]

There is also a strong element of predestiny, or predisposition, in the recent work on the so-called anthropic principle. Here the emphasis lies not with additional laws or organizing principles, but with the constants of physics. As we saw in Chapter 11, the values adopted by these constants are peculiarly felicitous for the eventual emergence of complex structures, and especially living organisms. Again, there is no compulsion. The

201

constants do not *determine* the subsequent structures, but they do *encourage* their appearance.

Predestiny is only a way of thinking about the world. It is not a scientific theory. It receives support, however, from those experiments that show how complexity and organization arise spontaneously and naturally under a wide range of conditions. I hope the review given in this book will have convinced the reader of the unexpectedly rich possibilities for self-organization that are being discovered in recent research.

There is always the hope that a really spectacular discovery will affirm the predestinist line of thinking. If life were discovered elsewhere in the universe, or created in a test tube, it would provide powerful evidence that there are creative forces at work in matter that encourage it to develop life; not vital forces or metaphysical principles, but qualities of self-organization that are not contained in – or at least do not obviously follow from – our existing laws of physics.

What does it all mean?

I should like to finish by returning to the point made at the beginning of this chapter. If one accepts predisposition in nature, what does that have to say about meaning and purpose in the universe?

Many people will find in the predestinist position support for a belief that there is indeed a cosmic blueprint, that the present nature of things, including the existence of human beings, and maybe even each particular human being, is part of a preconceived plan designed by an all-powerful deity. The purpose of the plan and the nature of the end state will obviously remain a matter of personal preference.

Others find this idea as unappealing as determinism. A plan that rigidly legislates the detailed course of human and non-human destiny seems to them a pointless charade. If the end state is part of the design, they ask, why bother with the construction phase at all? An all-powerful deity would be able to simply create the finished product at the outset.

A third point of view is that there is no detailed blueprint, only a set of laws with an inbuilt facility for making interesting things happen. The universe is then free to create itself as it goes along. The general pattern of development is 'predestined', but the details are not. Thus, the existence of intelligent life at some stage is inevitable; it is, so to speak, written into the laws of nature. But man as such is far from preordained.

Critics of predisposition dislike the anthropocentrism to which it seems to lead, but the requirement that the universe merely become self-aware at some stage seems a very weak form of anthropocentrism. Yet the knowledge that our presence in the universe represents a *fundamental* rather than an *incidental* feature of existence offers, I believe, a deep and satisfying basis for human dignity.

In this book I have taken the position that the universe can be understood by the application of the scientific method. While emphasizing the shortcomings of a purely reductionist view of nature, I intended that the gaps left by the inadequacies of reductionist thinking should be filled by additional scientific theories that concern the collective and organizational properties of complex systems, and not by appeal to mystical or transcendent principles. No doubt this will disappoint those who take comfort in the failings of science and use any scientific dissent as an opportunity to bolster their own anti-scientific beliefs.

I have been at pains to argue that the organizational principles needed to supplement the laws of physics are likely to be forthcoming as a result of new approaches to research and new ways of looking at complexity in nature. I believe that science is in principle able to explain the existence of complexity and organization at all levels, including human consciousness, though only by embracing the 'higher-level' laws. Such a belief might be regarded as denying a god, or a purpose in this wonderful creative universe we inhabit.

I do not see it that way. The very fact that the universe *is* creative, and that the laws have permitted complex structures to emerge and develop to the point of consciousness – in other words, that the universe has organized its own self-awareness – is for me powerful evidence that there is 'something going on' behind it all. The impression of design is overwhelming. Science may explain all the processes whereby the universe evolves its own destiny, but that still leaves room for there to be a meaning behind existence.

References

1 Blueprint for a universe

1 Ilya Prigogine, 'The Rediscovery of Time', in Sara Nash (ed.), *Science and Complexity* (Northwood, Middlesex, Science Reviews Ltd, 1985), p. 11.
2 Karl Popper and John Eccles, *The Self and Its Brain* (Berlin, Springer International, 1977), p. 61.
3 Ilya Prigogine and Isabelle Stengers, *Order Out of Chaos* (London, Heinemann, 1984), p. 9.
4 Erich Jantsch, *The Self-Organizing Universe* (Oxford, Pergamon, 1980), p. 96.
5 Louise B. Young, *The Unfinished Universe* (New York, Simon & Schuster, 1986), p. 15.

2 The missing arrow

1 P. S. Laplace, *A Philosophical Essay on Probabilities* (New York, Dover, 1951, original publication 1819), p. 4.
2 Richard Wolkomir, 'Quark City', *Omni*, February 1984, p. 41.
3 B. Russell, *Why I am not a Christian* (New York, Allen & Unwin, 1957), p. 107.
4 F. Engels, *Dialectics of Nature* (London, Lawrence & Wishart, 1940), p. 23.
5 A. S. Eddington, *The Nature of the Physical World* (Cambridge University Press, 1928), p. 74.

3 Complexity

1 H. Bergson, *Creative Evolution*, trans. A. Mitchell (London, Macmillan, 1964), p. 255.
2 P. Cvitanovic (ed.), *Universality in Chaos* (Bristol, Adam Hilger, 1984), p. 4.
3 Joseph Ford, 'How random is a coin toss?', *Physics Today*, April 1983, p. 4.

4 Chaos

1 Genesis 41:15.
2 D. J. Tritton, 'Ordered and chaotic motion of a forced spherical pendulum', *European Journal of Physics*, 7, 1986, p. 162.
3 Henri Poincaré, *Science and Method*, translated by Francis Maitland (London, Thomas Nelson, 1914), p. 68.
4 Ilya Prigogine, *From Being to Becoming: Time and Complexity in the Physical Sciences* (San Francisco, Freeman, 1980), p. 214.
5 Ford, op. cit.

5 Charting the irregular

1 P. S. Stevens, *Patterns in Nature* (Boston, Little, Brown, 1974).
2 D'Arcy W. Thompson, *On Growth and Form* (Cambridge University Press, 1942).
3 Stephen Wolfram, 'Statistical mechanics of cellular automata', *Reviews of Modern Physics*, 55, 1983, p. 601.
4 Oliver Martin, Andrew M. Odlyzko and Stephen Wolfram, 'Algebraic properties of cellular automata', *Communications in Mathematical Physics*, 93, 1984, p. 219.
5 Ibid. p. 221.
6 J. von Neumann, *Theory of Self-Reproducing Automata*, A. W. Burks (ed.) (Urbana, University of Illinois Press, 1966).
7 James P. Crutchfield, 'Space-time dynamics in video feedback', *Physica*, 10D, 1984, p. 229.
8 Wolfram, op. cit., p. 601.

References

6 Self-organization

1 Prigogine and Stengers, op. cit., Foreword by Alvin Tofler, p. xvi.
2 Charles H. Bennett, 'On the nature and origin of complexity in discrete, homogeneous, locally-acting systems,' *Foundations of Physics*, 16, 1986, p. 585.
3 Prigogine, 1980, op. cit., p. 147.

7 Life: its nature

1 James Lovelock, *Nature*, 320, 1986, p. 646.
2 C. Bernard, *Leçons sur les phénomènes de la vie*, 2nd edn (Paris, J. B. Baillière, 1885), vol. 1.
3 Jacques Monod, *Chance and Necessity* (London, Collins, 1972), p. 20.
4 Ibid., p. 31.
5 G. Montalenti, 'From Aristotle to Democritus via Darwin', in Francisco Jose Ayala and Theodosius Dobzhansky (eds), *Studies in the Philosophy of Biology* (London, Macmillan, 1974), p. 3.
6 H. H. Pattee, 'The Physical basis of coding', in C. H. Waddington (ed.), *Towards a Theoretical Biology* (4 vols, Edinburgh University Press, 1968), vol. 1, p. 67.
7 James P. Crutchfield, J. Doyne Farmer, Norman H. Packard and Robert Shaw, 'Chaos', *Scientific American*, December 1986, p. 38.

8 Life: its origin and evolution

1 Fred Hoyle, *The Intelligent Universe* (London, Michael Joseph, 1983).
2 J. Maynard-Smith, 'The status of neo-Darwinism', in C. H. Waddington (ed.), *Towards a Theoretical Biology* (4 vols, Edinburgh University Press, 1969), vol. 2, p. 82.
3 Motoo Kimura, *The Neutral Theory of Molecular Evolution* (Cambridge University Press, 1983).
4 S. J. Gould and N. Eldridge, *Paleobiology*, 3, 1977, p. 115.
5 Stuart A. Kaufmann, 'Emergent properties in random complex automata', *Physica*, 10D, 1984, p. 145.
6 Jantsch, op. cit., p. 101.

References

9 The unfolding universe

1 John D. Barrow and Joseph Silk, *The Left Hand of Creation* (New York, Basic Books, 1983), p. ix.

10 The source of creation

1 Bergson, op. cit., p. 41.
2 Popper and Eccles, op. cit., p. 14.
3 Kenneth Denbigh, *An Inventive Universe* (London, Hutchinson, 1975), p. 145.
4 Ibid., p. 147.
5 Arthur Peacocke, *God and the New Biology* (London, J. M. Dent, 1986).
6 William H. Thorpe, 'Reductionism in biology', in Francisco Jose Ayala and Theodosius Dobzhansky (eds), *Studies in the Philosophy of Biology* (London, Macmillan, 1974), p. 109.
7 P. W. Anderson, *Science*, 177, 1972, p. 393.
8 Bernhard Rensch, 'Polynomistic determination of biological processes', in Fransisco Jose Ayala and Theodosius Dobzhansky (eds), *Studies in the Philosophy of Biology* (London, Macmillan, 1974), p. 241.
9 A. I. Oparin, *Life, Its Nature, Origin and Development*, trans. A. Synge (New York, Academic Press, 1964).
10 Peter Medawar, 'A geometric model of reduction and emergence', in Fransisco Jose Alaya and Theodosius Dobzhansky (eds), *Studies in the Philosophy of Biology* (London, Macmillan, 1974), p. 57.
11 Montalenti, op. cit., p. 13.
12 Peacocke, op. cit.
13 Walter M. Elsasser, *Atom and Organism* (Princeton University Press, 1966), pp. 4 and 45.
14 E. P. Wigner, 'The probability of the existence of a self-reproducing unit', in *The Logic of Personal Knowledge*, anonymously edited (London, Routledge & Kegan Paul, 1961), p. 231.
15 Ibid.
16 John D. Barrow and Frank Tipler, *The Cosmological Anthropic Principle* (Oxford University Press, 1986), p. 237.
17 H. H. Pattee, 'The problem of biological hierarchy', in C. H. Waddington (ed.), *Towards a Theoretical Biology* (4 vols, Edinburgh University Press, 1970), vol. 3, p. 117.

References

18 Donald T. Campbell, '"Downward Causation", in hierarchically organized biological systems', in Francisco Jose Ayala and Theodosius Dobzhansky (eds), *Studies in the Philosophy of Biology* (London, Macmillan, 1974), p. 179.

19 N. Wiener, *Cybernetics* (Cambridge, Mass., MIT Press, 1961); E. M. Dewan, 'Consciousness as an emergent causal agent in the context of control system theory', in Gordon G. Globus, Grover Maxwell and Irwin Savodnik (eds), *Consciousness and the Brain* (New York and London, Plenum, 1976), p. 181.

11 Organizing principles

1 Prigogine, op. cit., p. 23.
2 Prigogine and Stengers, op. cit., pp. 285–6.
3 Daivd Bohm, 'Some remarks on the notion of order', in C. H. Waddington (ed.) *Towards a Theoretical Biology* (4 vols, Edinburgh University Press, 1969), vol. 2, p. 18.
4 Robert Rosen, 'Some epistemological issues in physics and biology', in B. J. Hiley and F. D. Peat (eds), *Quantum Implications: Essays in Honour of David Bohm* (London, Routledge & Kegan Paul, 1987).
5 Ibid.
6 C. G. Jung, 'Synchronicity: an acausal connecting principle', in *The Structure and Dynamics of the Psyche*, trans. R. F. C. Hull (London, Routledge & Kegan Paul, 1960), p. 423.
7 Ibid., p. 423.
8 Ibid., p. 530.
9 Arthur Koestler, *The Roots of Coincidence* (London, Hutchinson, 1972).
10 Rupert Sheldrake, *A New Science of Life* (London, Blond & Briggs, 1981).

12 The quantum factor

1 J. von Neumann, *Mathematical Foundations of Quantum Mechanics* (Princeton University Press, 1955).
2 E. Wigner, 'Remarks on the mind-body question', in I. J. Good (ed.), *The Scientist Speculates* (London, Heinemann, 1961), pp. 288–9
3 E. Wigner, 'The probability of the existence of a self-reproducing unit', op. cit.

References

4 Werner Heisenberg, *Physics and Beyond* (London, Allen & Unwin, 1971), p. 113.

5 Roger Penrose, 'Big bangs, black holes and "time's arrow"', in Raymond Flood and Michael Lockwood (eds), *The Nature of Time* (Oxford, Blackwell, 1986).

6 J. A. Wheeler, 'Bits, quanta, meaning', in A. Giovannini, M. Marinaro and A. Rimini (eds), *Problems in Theoretical Physics* (University of Salerno Press, 1984) p. 121.

7 C. O. Alley, O. Jakubowicz, C. A. Steggerda and W. C. Wickes, 'A delayed random choice quantum mechanics experiment with light quanta', in S. Kamefuchi et al. (eds) *Proceedings of the International Symposium on the Foundations of Quantum Mechanics, Tokyo 1983*, (Tokyo, Physical Society of Japan, 1984) p. 158.

8 Prigogine, 1980, op. cit., p. 199.

9 Eddington, op. cit., p. 98.

10 David Bohm, *Wholeness and the Implicate Order* (London, Routledge & Kegan Paul, 1980).

11 A. Aspect et al., *Physical Review Letters*, **49**, 1982, p. 1804.

12 J. S. Bell, *Physics*, **1**, 1964, p. 195.

13 H. H. Pattee, 'Can life explain quantum mechanics?', in Ted Bastin (ed.), *Quantum Theory and Beyond* (Cambridge University Press, 1971).

14 W. M. Elsasser, 'Individuality in biological theory', in C. H. Waddington (ed.), *Towards a Theoretical Biology* (4 vols, Edinburgh University Press, 1970), vol. 3, p. 153.

15 Erwin Schrödinger, *What is Life?* (Cambridge University Press, 1944), p. 76.

16 William T. Scott, *Erwin Schrödinger: An Introduction to his Writings* (Amherst, University of Massachusetts Press, 1967).

17 N. Bohr, *Nature*, 1 April 1933, p. 458.

18 Heisenberg, op. cit., ch. 9.

13 Mind and brain

1 J. J. Hopfield, D. I. Feinstein and R. G. Palmer, '"Unlearning" has a stabilizing effect in collective memories', *Nature*, **304**, 1983, p. 158.

2 Crutchfield et al., op. cit., p. 49.

3 Freeman Dyson, *Disturbing the Universe* (New York, Harper & Row, 1979), p. 249.

References

4 R. W. Sperry, 'Mental phenomena as causal determinants in brain function', in Gordon Globus, Grover Maxwell and Irwin Savodnik (eds), *Consciousness and the Brain* (New York and London, Plenum, 1976), p. 166.
5 Ibid., p. 165.
6 Ibid., p. 167.
7 Roger Sperry, *Science and Moral Priority* (New York, Columbia University Press, 1983), p. 92.
8 D. M. Mackay, *Nature*, 232, 1986, p. 679.
9 Marvin Minsky, *The Society of Mind* (New York, Simon & Schuster, 1987), p. 26.
10 Michael Polanyi, 'Life's irreducible structure', *Science*, 1968, p. 1308.

14 Is there a blueprint?

1 J. B. Hartle and S. W. Hawking, 'Wave function of the universe', *Physical Review*, D 28, 1983, p. 2960.

Further Reading

ABRAHAM, Ralph H. and SHAW, Christopher D. *Dynamics: The Geometry of Behaviour*, three volumes (Aerial Press: Santa Cruz, 1982).

AYALA, Francisco J. and DOBZHANSKY, Theodosius (eds). *Studies in the Philosophy of Biology* (Macmillan: London, 1974).

BARROW, John and SILK, Joseph. *The Left Hand of Creation* (Basic Books: New York, 1983).

BARROW, John and TIPLER, Frank. *The Cosmological Anthropic Principle* (Oxford University Press: Oxford, 1986).

BOHM, David. *Wholeness and the Implicate Order* (Routledge & Kegan Paul, 1980).

CLARK, John W., WINSTON, Jeffrey V. and RAFELSKI, Johann. 'Self-organization of neural networks', *Physics Letters*, 102A, 1984, p. 207.

CRICK, Francis. *Life Itself: Its Origin and Nature* (Simon & Schuster, New York, 1982).

CVITANOVIC, Predrag (ed.). *Universality in Chaos* (Adam Hilger: Bristol, 1984).

DAVIES, Paul. *Superforce* (Heinemann: London/Simon & Schuster: New York, 1984).

DAVIES, P. C. W. and BROWN, J. R. (eds). *The Ghost in the Atom* (Cambridge University Press: Cambridge, 1986).

DAWKINS, Richard. *The Blind Watchmaker* (Longman: London, 1986).

DENBIGH, Kenneth. *An Inventive Universe* (Hutchinson: London, 1975).

DEWDNEY, A. K. 'Exploring the Mandelbrot set', *Scientific American*, August 1985, p. 8.

EIGEN, M. and SCHUSTER. P. *The Hypercycle* (Springer-Verlag: Heidelberg, 1979).

ELSASSER, Walter M. *Atom and Organism* (Princeton University Press: Princeton, 1966).

Further Reading

FARMER, Doyne, TOFFOL, Tommaso and WOLFRAM, Stephen (eds). 'Cellular automata', *Physica*, 10D (North-Holland: Amsterdam, 1984).

FARMER, Doyne, TOFFOL, Tommaso and WOLFRAM, Stephen (eds). 'Evolution, games and learning: models for adaptation in machines and nature', *Physica*, 22D (North Holland: Amsterdam, 1986).

FORD, Joseph. 'How random is a coin toss?' *Physics Today*, April 1983, p. 40.

FORD, Joseph. 'What is chaos, that we should be mindful of it?' In Davies, P. C. W. (ed.), *The New Physics* (Cambridge University Press: Cambridge, 1988).

GEHRING, Walter J. 'The molecular basis of development', *Scientific American*, October 1985, p. 137.

GLEICK, James. *Chaos: Making a New Science* (Viking: New York, 1987).

GLOBUS, Gordon G., MAXWELL, Grover and SAVODNIK, Irwin. *Consciousness and the Brain* (Plenum: New York and London, 1976).

GREENBERG, Richard and BRAHIC, André (eds). *Planetary Rings* (The University of Arizona Press: Tucson, 1984).

GRIBBIN, John. *In Search of the Big Bang* (Heinemann: London, 1986).

HAKEN, H. *Synergetics* (Springer-Verlag: Berlin, 1978).

HINTON, Geoffrey E. 'Learning in parallel networks', *Artificial Intelligence*, April 1985, p. 265.

JANTSCH, Erich. *The Self-Organizing Universe* (Pergamon: Oxford, 1980).

KADANOFF, Leo P. 'Roads to chaos', *Physics Today*, December 1983, p. 46.

MANDELBROT, Benoit B. *The Fractal Geometry of Nature* (Freeman: San Francisco, 1982).

MAYNARD-SMITH, John. *The Theory of Evolution* (Penguin Books: Harmondsworth, Middx., 3rd edn, 1975).

MONOD, Jacques. *Chance and Necessity* (Collins: London, 1972).

MORRIS, Richard. *Time's Arrows* (Simon & Schuster: New York, 1984).

NELSON, David R. 'Quasicrystals', *Scientific American*, August 1986, p. 42.

NICOLIS, G. 'Physics of far-from-equilibrium systems and self-organization'. In P. C. W. Davies (ed.) *The New Physics* (Cambridge University Press: Cambridge, to be published 1988).

NICOLIS, G. and PRIGOGINE, I. *Self-Organization in Non-Equilibrium Systems* (Wiley: New York, 1977).

PAGELS, Heinz. *Perfect Symmetry* (Michael Joseph: London, 1985).

PENROSE, Roger. 'Big bangs, black holes and "time's arrow"'. In B. J. Hiley and F. D. Peat (eds), *Quantum Implications: Essays in Honour of David Bohm* (London, Routledge & Kegan Paul, 1987).

POLANYI, Michael. 'Life's irreducible structure', *Science*, June 1968, p. 1308.

POPPER, Karl R. *The Open Universe: An Argument for Indeterminism* (Hutchinson: London, 1982).

POPPER, Karl R. and ECCLES, John C. *The Self and its Brain* (Springer International: Berlin, 1977).

PRIGOGINE, Ilya. *From Being to Becoming: Time and Complexity in the Physical Sciences* (Freeman: San Francisco, 1980).

PRIGOGINE, Ilya and STENGERS, Isobelle. *Order Out of Chaos* (Heinemann: London, 1984).

RAE, Alastair. *Quantum Physics: Illusion or Reality* (Cambridge University Press: Cambridge, 1986).

RICHTER, P. H. and PEITGEN, H. O. *The Beauty of Fractals* (Springer: Berlin and New York, 1985).

ROSEN, Robert (ed.). *Theoretical Biology and Complexity* (Academic Press: New York, 1985).

ROSEN, Robert. *Anticipatory Systems* (Pergamon: London, 1986).

SANDER, L. M. 'Fractal growth processes', *Nature*, 28 August 1986, p. 789.

SCHRÖDINGER, Erwin. *What is Life?* (Cambridge University Press: Cambridge, 1944).

SCHUSTER, Heinz G. *Deterministic Chaos* (Physik-Verlag GmbH: Weinheim, Germany, 1984).

SHAPIRO, Robert. *Origins: A Skeptic's Guide to the Creation of Life on Earth* (Summit Books: New York, 1986).

THOM, R. *Structural Stability and Morphogenesis* (Benjamin: Reading, Mass. 1975).

THOMPSON, D'Arcy W. *On Growth and Form* (Cambridge University Press: Cambridge, 1942).

THOMPSON, J. M. T. and STEWART, H. B. *Nonlinear Dynamics and Chaos* (Wiley: New York, 1986).

TURING, A. M. 'The chemical basis of morphogenesis', *Philosophical Transactions of the Royal Society*, B237, 1952, p. 37.

WADDINGTON, C. H. (ed.). *Towards a Theoretical Biology*, four volumes (Edinburgh University Press: Edinburgh, 1968–72).

WHEELER, J. A. 'Bits, quanta, meaning', in A. Giovannini, M. Marinaro, F. Mancini and A. Rimini (eds) *Problems in Theoretical Physics* (University of Salerno Press, 1984).

WOLPERT L. 'Pattern formation in biological development', *Scientific American*, 239 154, 1978.

YOUNG, Louise B. *The Unfinished Universe* (Simon & Schuster: New York, 1986).

Index

Index

Index